The Transition to
Agile Manufacturing

Also available from ASQC Quality Press

Business Process Benchmarking: Finding and Implementing
Best Practices
Robert C. Camp

Actual Experiences of a CEO: How to Make Continuous Improvement
in Manufacturing Succeed for Your Company
Hank McHale

Small Business Success Through TQM: Practical Methods to Improve
Your Organization's Performance
Terry Ehresman

Performance-Based Assessments: External, Internal, and
Self-Assessment Tools for Total Quality Management
Paul F. Wilson and Richard D. Pearson

Linking Quality to Profits: Quality-Based Cost Management
Hawley Atkinson, John Hamburg, and Christopher Ittner

Management in a Quality Environment
David Griffiths

Reengineering the Factory: A Primer for World-Class Manufacturing
A. Richard Shores

To request a complimentary catalog of publications, call 800-248-1946.

The Transition to Agile Manufacturing

Staying Flexible for Competitive Advantage

Joseph C. Montgomery
Lawrence O. Levine
editors

ASQC Quality Press
Milwaukee, Wisconsin

The Transition to Agile Manufacturing: Staying Flexible for Competitive Advantage
Joseph C. Montgomery and Lawrence O. Levine, editors

Library of Congress Cataloging-in-Publication Data

The transition to agile manufacturing: staying flexible for
  competitive advantage / Joseph C. Montgomery, Lawrence O. Levine,
  editors.
      p. cm.
  Includes bibliographical references and index.
  ISBN 0-87389-347-6
  1. Flexible manufacturing systems.   2. Production engineering.
3. Industrial management.   I. Montgomery, Joseph C. II. Levine,
Lawrence O., 1954– .
TS155.65.T73   1995
658.5 — dc20   95-42622
                                          CIP

©1996 by ASQC

Permission Acknowledgments
The editors, Joseph C. Montgomery and Lawrence O. Levine, prepared the material presented in this book while staff members at Pacific Northwest National Laboratory, operated for the U.S. Department of Energy by Battelle Memorial Institute under contract DE-AC06-76RLO 1830.

Pages 36–40, "Difficulties of Change Due to System Complexity": From *The Fifth Discipline* by Peter M. Senge. Used by permission of Doubleday, a division of Bantam Doubleday Dell Publishing Group, Inc.

Pages 58–66, "A Case Study of Successful Management of Change": Summary of Xerox changes from *Prophets in the Dark* by David T. Kearns and David A. Nadler. Copyright © 1992 David T. Kearns and David A. Nadler. Reprinted by permission of HarperCollins Publishers, Inc.

Pages 247–48, quote from Walt Disney: Used by permission from The Walt Disney Company.

10 9 8 7 6 5 4 3 2 1

ISBN 0-87389-347-6

Acquisitions Editor: Susan Westergard
Project Editor: Jeanne W. Bohn

ASQC Mission: To facilitate continuous improvement and increase customer satisfaction by identifying, communicating, and promoting the use of quality principles, concepts, and technologies; and thereby be recognized throughout the world as the leading authority on, and champion for, quality.

Attention: Schools and Corporations
ASQC Quality Press books, audiotapes, videotapes, and software are available at quantity discounts with bulk purchases for business, educational, or instructional use. For information, please contact ASQC Quality Press at 800-248-1946, or write to ASQC Quality Press, P.O. Box 3005, Milwaukee, WI 53201-3005.

For a free copy of the ASQC Quality Press Publications Catalog, including ASQC membership information, call 800-248-1946.

Printed in the United States of America

 Printed on acid-free paper

 ASQC
Quality Press
611 East Wisconsin Avenue
Milwaukee, Wisconsin 53202

Contents

Chapter 2. Managing Systemwide Change

Chapter 5. Agile Practices

Chapter 7. Strategic Direction

Chapter 8. Performance Measures

Foreword

In 1988, with the passage of the Omnibus Trade and Competitiveness Act, the National Institute of Standards and Technology (NIST) embarked on an exciting and challenging new initiative called the Manufacturing Technology Centers (MTC) program. The MTC program (now known as the Manufacturing Extension Partnership, or MEP) was tasked with the specific mission of improving the productivity and technological performance of U.S.-based manufacturing firms, with a particular emphasis on small and medium-sized manufacturing establishments. The underlying philosophy of the program was that within NIST, as well as within many other federal laboratories, one could readily find the latest, most advanced manufacturing technologies, and these technologies could simply be transferred to small firms to improve their competitive ability in the global marketplace. Local Manufacturing Technology Centers were created by NIST to act as intermediaries or transfer points between the federal laboratories and the small firms they were seeking to assist. A classic case of the proverbial hammer looking for a nail, NIST and its technology centers set out with a treasure chest of factory automation protocols, seven-axis robotic systems, feature-based design, and other leading-edge technologies in search of small firms ready, willing, and able to adopt these latest and greatest techniques. But such customers were very few and very far between. By design, the program was destined for failure.

We needed to learn some hard lessons in those early years. Among them was the fact that while NIST and the other federal laboratories offered advanced technology, and small manufacturing firms needed technology, *what we had wasn't necessarily what they needed.* Many of the firms with which the earlier centers began to work were several generations behind the state-of-the-market, much less the state-of-the-art. (Unfortunately, this remains true for the vast majority of small manufacturing firms today.) The first, and perhaps single most important, lesson learned was *to succeed, we need to be market driven.* We needed to listen to our customers and fashion solutions unique to their particular circumstances. This meant delivering the most appropriate technologies in lieu of the most advanced ones. As well, this meant that no two company-specific solutions would be alike. And, in the greatest sense of the word, success rested largely on our ability to become and remain an agile service provider.

Our next greatest discovery was that *technology, in and of itself, is not a solution:* at least not technology as narrowly defined for the purposes of our program. We focused initially on what could conveniently be categorized single-point hardware or software-type solutions—the machines, materials, processes, production floor (the trees)—with little or no regard for what the authors refer to as the manufacturing production system (the forest). In the book *Empowering Technology: Implementing a U.S. Strategy,* Lewis M. Branscomb defines technology as "the aggregation of capabilities, facilities, skills, knowledge, and organization required to successfully create a useful service or product." It is with this broader, yet more appropriate, definition of technology that we began to understand the nature and scope of technical services that the technology centers needed to deliver to small firms. One cannot attempt to implement any new tools or techniques in the production environment without duly considering the effect that change has on the greater manufacturing production system. This is how we found ourselves evolving from a hard technology–oriented program in the late 1980s to today's partnership among several federal agencies, more than 40 technology centers in 31 states, and more than

700 service provider organizations nationwide to bring together an integrated service delivery system that can address and assist in areas of technology adoption, workforce development and workplace organization, financing, marketing, quality, enterprise integration, and environmental services, to name a few.

Mr. Levine, Dr. Montgomery, and the other contributing authors to *The Transition to Agile Manufacturing: Staying Flexible for Competitive Advantage* have compiled a most comprehensive road map for small manufacturing firms that addresses the issue of greater production and business flexibility necessary to meet the increasingly competitive demands of the global marketplace. The treatment of the whole business as an integrated manufacturing production system is not only oriented toward those firms seeking expansion and growth but is an unavoidable requirement for those wanting to continue participating.

Finally, I'm reminded of a visit I made nearly a year ago to one of the MEP extension centers in the midwest. One of our best and most enthusiastic field agents boasted to me of how he would drag reluctant companies kicking and scratching into the process of modernization. A silent and confusing haze fell across a room filled with our colleagues when I asked, "Why?" We, at our very best, can only lead companies to the "waters of change," the hard part, their part, is making the investment necessary to implement the change. The authors of *The Transition to Agile Manufacturing* have built for small manufacturers a road map into the world of agile manufacturing, a world that will define manufacturing and the competitive environment of the future. Mass production is soon to be replaced by mass customization in a marketplace that is growing increasingly more demanding in terms of cost and quality and increasingly less patient in terms of time.

Kevin Carr
Director, Manufacturing Extension Partnership
National Institute of Standards and Technology

Preface

The requirements for remaining competitive in manufacturing keep getting higher—there seems to be no end in sight. Only recently, high quality and efficiency were the "necessary and sufficient" conditions for staying in business. Now, however, manufacturers must be able to rapidly develop and produce customized products to meet customer needs. To further complicate matters, the requirements for economies of scale, based on the traditional assumptions of mass production, are coming into direct conflict with the requirements for economies of scope—that is, maintaining continuous innovation while using people and equipment to cost-effectively produce smaller amounts of a range of products. *The Transition to Agile Manufacturing* describes how companies—small to medium-sized manufacturing companies—can mobilize their resources to remain competitive.

Because agile manufacturing has been described in several different ways in the literature, let us state the basic assumptions upon which this book is based.

• First, agile manufacturing is built on a foundation of some, but not all, of the practices common to lean manufacturing (Womack, Jones, and Roos 1990). For example, while both lean and agile manufacturing emphasize small batch sizes, agile manufacturing takes the additional steps of reducing product development time and allowing for considerable customization of product features. Because of the common foundations of agile and lean manufacturing, the old, yet effective, requirements for implementing lean manufacturing can generally be redirected, rather than scrapped.

• Second, agility requires careful integration of people, technology, and organization/business elements. We refer to this integration as *alignment* to indicate that all parts of the organization are working toward a common goal. The barriers to achieving alignment are as much rooted in people and organizational issues, as well as the difficulties of managing change, as they are technical manufacturing problems. Therefore, transforming an organization to agile practices requires the talents of some experts from outside the field of manufacturing—including those familiar with organizational design and management of the change process.

• Third, we do not hold the same view as those proponents of agile manufacturing who stress the importance of different firms communicating and coordinating with one another via sophisticated electronic systems to manage integrated supply chains. This form of organization, labeled the "virtual organization" (Iacocca Institute 1991; Davidow and Malone 1992), is premature at this time. Many issues remain unresolved regarding the management of virtual organizations, such as the appropriate distribution of profits, coordination and finance of investments across firms, and assumption of responsibility for product liability. These issues address building and maintaining trust among partners and customers, rather than the more technical issues of developing and implementing common standards for electronic data interchange and coordination. Because this book is focused on the needs of small and mid-sized manufacturers, we think it prudent for them to focus on getting their own manufacturing, product development, and business processes in order first. They will then be prepared for participating in virtual organizations, if and when such organizations become yet a new requirement for success. At this time, agility must rest firmly on flexibility in business processes and on the capabilities of the shop floor.

The key themes of this book are agility, alignment, and change management. In order to address these themes, we have written a somewhat unusual book. By relying on a number of different authors for the chapters, we have tried to capture the variety of knowledge and skills necessary to make a successful transition to agile manufacturing. The result, we believe, is more than just an edited volume of readings. Here are some of the main perspectives we have tried to capture: (1) organizational and business management, (2) industrial and mechanical engineering, and (3) production and inventory management. Most books on manufacturing focus on just one of these perspectives. We have combined them because we feel all three are required to successfully transform and manage a successful manufacturing firm along the path to agility.

Books on manufacturing also vary in the degree to which they focus on theory versus practice. Given the intended audience for this book (general managers, production managers, and engineers in small to mid-sized companies) we have tried to combine both theory and practice. We have worked for just enough theory to explain the "whys" but have included enough practical information so that agile practices actually can be implemented by companies with limited resources.

Finally, many books on manufacturing excellence focus on large companies—those with large internal technical resources and/or the ability to bring in expensive outside consultants and equipment. This book provides practical support for companies of smaller size who realistically have limited resources with which to implement agile practices. In addition, we have included references for low-cost assistance available to smaller companies as well as references to additional sources of technical information. Consequently, the reader will encounter more references than are found in some of the more popular books on manufacturing, but less than in a more strictly academic work.

The authors of this book, with the exception of Dr. Lemak and Dr. Paul, are employed by Battelle Pacific Northwest National Laboratories, where they conduct research and are consultants to public- and private-sector organizations. Battelle Pacific Northwest National Laboratories is a major research component of Battelle Memorial Institute, which is the world's largest and oldest contract research organization. Since 1965, Battelle has operated Pacific Northwest National Laboratory (PNNL) in Richland, Washington, for the U.S. Department of Energy (DOE). As a multiprogram national laboratory, PNNL provides extensive capabilities for conducting research and technology development for DOE, many other federal agencies, and private industry. Together, the extensive client base of Battelle and PNNL and the interdisciplinary approach to problem solving have given the authors of this book a unique depth and breadth of experience, which is reflected throughout this book.

The book is organized in three sections. Section one (preface, chapters 1 and 2) introduces the topic of agile manufacturing. This section provides an overview of agility (preface), an operational description of the successful agile production organization (chapter 1), and a discussion of the change management issues and strategies associated with a successful transformation to agile practices (chapter 2).

Section two (chapters 3–6) focuses on the more technical and engineering aspects of becoming agile. This section provides techniques to model/simulate the business processes (chapter 3), a methodology for redesigning the current work processes (chapter 4), a description of agile manufacturing practices (chapter 5), and guidelines for introducing new technology into the manufacturing process (chapter 6).

Section three (chapters 7–10) addresses additional business and organizational issues. These issues include the new "strategic identity" an agile organization takes on with markets, customers, and competitors (chapter 7); the need for customer-oriented performance measures to improve performance and overcome functional boundaries (chapter 8); the importance of ongoing organizational learning and skill acquisition (chapter 9); and the new role of management in agile production systems (chapter 10).

We believe that the topics covered in sections 1–3 form the foundation needed for creating an agile manufacturing system. The broad range of topics included in the chapters of this book may appear somewhat daunting to those who simply want to improve their manufacturing performance. However, no one person needs to become an expert on everything. Rather, those leading the change effort need to be selected to represent a diversity of backgrounds and skills and will need to then develop a general understanding of a broad range of issues. We hope and trust that these chapters will help guide you to make a successful transition in your company!

Acknowledgments

This book was originally conceived by Brian Paul. His enthusiasm and vision helped convince both the authors and Battelle management that there was a need for a book such as this and that we could help fill the need. Marye Hefty, former technical editor at PNNL, worked diligently with each author to help ensure that the book had a common tone, clarity, and focus on the intended audience. To the degree we have achieved these goals, Marye deserves much of the credit. The continued support and encouragement of our managers helped sustain this effort. Their confidence in us is much appreciated.

References

Davidow, W. H., and M. S. Malone. 1992. *The virtual corporation: Structuring and revitalizing the corporation for the twenty-first century.* New York: Harper Collins.

Iacocca Institute. 1991. *Twenty-first century manufacturing enterprise strategy. An industry-led view.* Vols. 1 and 2. Bethlehem, Pa.: Iacocca Institute.

Womack, J. P., D. T. Jones, and D. Roos. 1990. *The machine that changed the world.* New York: Rawson Associates.

CHAPTER 1

The Agile Production System
Joseph C. Montgomery

A new vision of manufacturing is needed to guide U.S. manufacturers to survival and success in the arena of global competition. Part of this vision involves the use of agile manufacturing, which provides the basis for highly adaptive, flexible, and efficient manufacturing practices, in the product development cycles as well as in the manufacturing cycle. Another key part of this vision is an organization that is tightly run and well coordinated, and that provides effective support for the efforts of the manufacturing system. This chapter will describe the functions of this organization, including the roles of marketing, engineering, human resources, purchasing, and production planning and control. All too often these organizations pursue their own agendas without a tightly coupled strategy for improving the overall performance of the company. These groups must be reoriented so that their entire mission and purpose is to provide support for effective product development and manufacturing cycles. They must become service organizations with an identified customer and a customer-service attitude. We will refer to manufacturing organizations that have accomplished this realignment as *task aligned* (from Beer, Eisenstat, and Spector 1990), and *alignment* as the process by which the reorientation is accomplished.

1

Figure 1.1 shows what is meant by the alignment process—how the production system must become the center of the organization with other functions providing service and support roles. Agile manufacturing combined with organization-wide task alignment are the necessary ingredients for a globally competitive company. Figure 1.1 provides our vision of the relationships between system functions as the manufacturing production system changes from a conventional to a task-aligned organization. The large boxes represent the entire organization during its transition to a new form. Each small box represents a system function, with the size of the box and its position (near the center or on the periphery) indicating its relative importance. Over time, production (shaded) evolves from being just "one of the crowd" to being the center of the system.

The right side of the figure shows the truly task-aligned organization. Here, production assumes the central, dominant role in the organization, with other functions aligned in appropriate supporting roles. This figure indicates the extent of change, the paradigm shift, that is necessary to achieve global competitiveness.

This figure helps convey a vision of the new production organization; however, it must be kept in mind that another aspect of the vision—the use of agile manufacturing—must be incorporated into the transition. Simply creating a task-aligned production system without agile manufacturing will greatly strengthen the existing organization, but will not be sufficient to establish global competitiveness.

In the next section, we will consider in more detail what is involved in agile manufacturing. Subsequent sections will address support subsystems and focus more on the need for alignment with production. As a convenient format for this discussion, we will refer to "conventional" approaches as compared to agile practices. The term *conventional practices* is used here to refer to those that were the standard 20 years or so ago and that, in many companies, continue on relatively unchanged today. Although most companies have improved significantly over the years and may be closer to agile practices than conventional ones in some areas, few have made systematic across-the-board improvements.

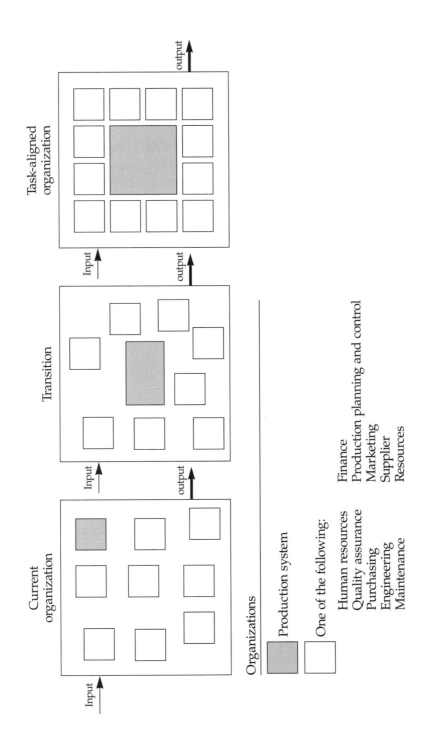

Figure 1.1. The task-aligned organization.

3

The Manufacturing Production System

As shown in Figure 1.2, manufacturing organizations are composed of three interrelated systems: production, production support, and business operations. Each component of these subsystems performs critical tasks and may have complex relationships and linkages with each other as well as with the business environment (customers, suppliers, regulators, and so on). The focal point of the organization is the production system, the function responsible for the output (highly reliable products) that satisfies customers' needs. Profits on sales of the output provide the input (finances, raw materials, resources) needed to resupply and sustain the entire system. Production cannot, of course, be accomplished without the help of the other two systems and their components. The overall organization, referred to as the manufacturing production system, emphasizes the interrelated nature of manufacturing and affirms the central, but highly dependent, position of production.

Production

Our adjective for describing the new production system is *agile,* meaning fast and flexible. Table 1.1 presents the key characteristics of the shop floor in the agile production system. Combined, these characteristics enable the production system to respond quickly with high-quality products to fluctuating customer demand and preference. Agile manufacturing originated with the revolutionary ideas of Toyota in the 1950s and 1960s and has been continually refined and improved since that time (Black 1991).

The goal of an agile production system is to quickly respond to customer demand with high-quality, low-cost products. This is accomplished through continuous process improvement: steadily eliminating waste, improving production processes, and rapidly introducing product innovations in response to market changes. Results of an agile production system include time to market compression, smaller economic production runs for products, and expanded product mixes.

Shop floor layout in an agile production system is product-oriented, resulting in simplified material flow and reduced material handling. Because of the close proximity of all processes and operations for a

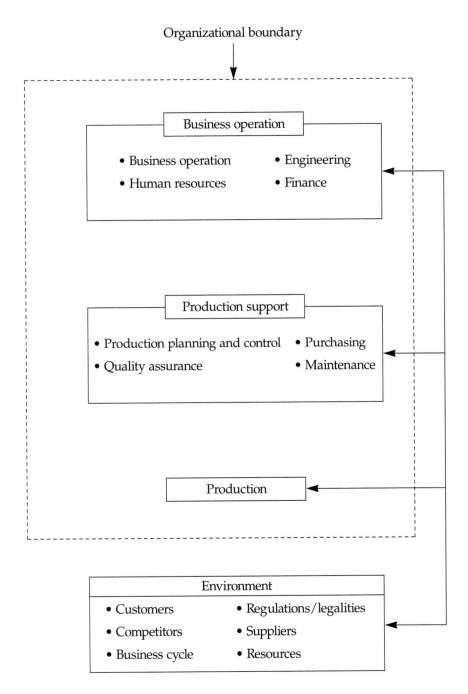

Figure 1.2. The manufacturing production system.

Table 1.1. Components of the agile manufacturing system.

Component	Description
Small batch size	Maintain small production runs.
Minimal buffer stock	Reduce buffer inventories to expose system problems.
Total quality control	Catch and correct errors at the source; avoid final inspections. Workers assume responsibility for quality.
Elimination of waste	Dispense with any activities not directly related to production. Use the minimum amount of time, equipment, parts, space, tools, and so on that add value to the product.
Setup reduction	Reduce work that must be done when machinery is stopped. Eliminate adjustments, simplify attachment and detachment. Train and practice to minimize time requirements.
Redesign of workflow	Adopt a product-oriented, rather than a process-oriented layout. Eliminate unnecessary transportation, work-in-process buffers, multiple handling of materials.
Improved work processes	Adopt cell manufacturing and statistical process control. Analyze and improve process routes. Obtain worker ideas for continuing improvements.
Visual control	Adopt line stop systems, trouble lights, production control boards, foolproof mechanisms, control charts.
Preventive maintenance	Have operators perform routine repairs and maintenance. Have maintenance staff support operators and perform difficult maintenance and repair.
Leveled/mixed production	Maintain steady rate of output using different product mix.
Reduced cycle time	Balance operator time utilization, reduce time needed to complete product.
Kanban system	Use kanban cards to pull products through system.
Continuous improvement	Employees find better ways to improve work processes.

given product, production control can be simplified and integrated into the shop floor using visual production signals *(kanbans)*. In addition, some of the work of other support functions, such as quality inspection and preventive/routine maintenance, can be incorporated into the tasks of the production workers in order to maximize these workers' control of production. (These two approaches are known as *quality at the source* and *total preventive maintenance*). The shift of these support tasks over to the production workers encourages continuous quality improvement, reduces the number of unanticipated work stoppages, and results in overall lower production costs.

As a result of organizing by product, workers are joined into teams that assume virtually full responsibility for the production process. Each team sees its own production efforts from the beginning to the end of the process and is well-positioned to work to improve the entire system. Common improvements include setup time reduction, use of statistical process control to improve quality and work processes, elimination of waste, and cross-training and job rotation so everyone sees the "big picture." These improvements in the production system result in a tightly coupled system of just-in-time material deliveries, high-quality production, and rapid response to problems and changing production demands. Further, the workers become highly trained and highly motivated by the responsibility, challenge, and decision-making ability that come with operating an agile production system.

In contrast, layouts in many conventional production systems are based on a process, or functional, orientation. For example, job shops typically locate similar machining processes or manual operations next to one another. As a result, workers in such areas are focused entirely on performing these specific functions and do not see a completed product.

Given a fragmented, poorly functioning conventional production system, management tends to be driven toward more centralized control, including creating a staff group to perform this function. Yet the centralized staff members do not have a clear picture of the production process to aid in isolating and solving production problems. Large amounts of inventory are pushed through the system in an attempt to compensate for quality and equipment problems, resulting in large work-in-process (WIP) buffers, storage problems, and bottlenecks. Ultimately, product quality and cost are sacrificed as management pushes to sustain production volumes. Process and product innovation are unlikely to thrive in such an environment.

With such centralized control in the conventional system, supervisors and workers on the shop floor focus on their own small portion of the process. Unfortunately, by doing so, without regard for the next step in the process, shop floor employees unintentionally suboptimize overall production performance (Schonberger 1990). These employees suboptimize the system of production because they do not provide materials to the next step exactly as they are needed, fail to get feedback from the next process regarding their performance, and do not attempt to coordinate with the previous step in the process.

In contrast, supervisors and teams in agile production systems are able to assume responsibility for the entire production process. Each part of the process for a given product is located so as to maximize communications and materials transfer with the steps before and after. Each step realizes and fulfills its role as the customer for the previous process and the supplier for the next step. The resulting chain of customers (see Schonberger 1990) means production is tightly linked from beginning to end. The employees can work to continuously improve the production process in order to optimize the overall performance of the production system.

Production Support

Because the layout and functioning of conventional production systems are often disjointed and inefficient, they are very hard to support. For example, in a line organized by process, scheduling and coordinating product flow involves routing of the products through multiple stations in multiple areas. Not only is this routing complicated, but trying to locate the source of a quality problem given all the alternative routes is a major undertaking. The maintenance function also has a difficult task. Because the conventional system relies on maintenance personnel for all forms of maintenance, preventive maintenance takes a back seat to urgent repair requests. With inadequate attention to proper preventive maintenance, maintenance personnel end up being overloaded

when asked to respond to the constant stream of emergency maintenance requests from supervisors throughout the entire plant.

Production support in an agile production system is greatly simplified because much of the work of production control, quality control,
and preventive maintenance is performed by the shop floor work
teams themselves. The need for detailed work center scheduling is virtually eliminated by the use of visual inventory control on the shop
floor. The work of quality assurance (QA) is greatly simplified because
of the straightforward product-oriented layout and because the work
teams assume responsibility for their own quality. The QA group may
be called in as-needed to help deal with difficult quality issues and to
provide needed training. Because shop floor personnel perform their
own preventive maintenance, more extensive maintenance can be performed by maintenance staff not overloaded with preventive maintenance and other basic tasks. The work of purchasing is simplified and
streamlined by greatly reducing the number of suppliers, establishing
long-term relationships with them, and working to improve the quality
of the supplies received.

The following four sections provide greater detail about the production support functions (production planning and control, QA, purchasing, and maintenance) in conventional versus agile manufacturing.
A common theme will be that, in an agile organization, each support
function is partially taken over by the workers of the production line,
leaving a reduced set of responsibilities to be performed by simplified
and downsized production support organizations.

Production Planning and Control

One of the key features of the conventional production, planning, and
control (PP&C) organization is the use of a production forecast, neces-
sitated by the existence of long production runs and long supplier lead
times. Days, weeks, or months may be needed to move inventory from
start to finish. As a result, the conventional factory needs large quantities of WIP and inventory in order to operate. Consequently, the PP&C
organization tries to schedule production to run as efficiently as possible. Complex, computer-driven material requirements planning (MRP)
systems may be used to set the quantities of materials and schedules

for material movement throughout the production line. Production orders are thus "pushed" through the system as workers at individual work locations strive to keep up with the schedule.

Unfortunately, PP&C planning is complicated by any number of issues including the following: (1) the number of production parts can be in the thousands (each with its own routing), (2) machines break down in an unpredictable fashion, (3) workers may be absent or make mistakes, and (4) orders may be in competition with one another to get through bottleneck resources. Each of these problems may result in the lines being shut down to wait for needed parts or repairs. Supervisors or workers may work frantically to try to locate needed parts. The PP&C organization tends to be blamed for all of these problems by production workers, resulting in an adversarial relationship between two key groups who should be working closely together.

In the agile production system, detailed scheduling is primarily the responsibility of the shop floor. Workers control production and inventory through the use of visual signals, or kanbans, in what is known as a "pull" manufacturing system. In the pull system, the production schedule is provided only to the final section of the production line. Work centers upstream from final assembly (including vendors) produce for downstream work centers based on the pull signal they receive. The pull system means there is no buildup of WIP and no need for complicated planning and scheduling activities. The pull system also drives upstream workstations to become adept at reducing changeover times, eliminating bottlenecks, reducing cycle times, and, in general, continuously improving the work processes.

Clearly, the role of PP&C in agile production is dramatically simplified and reduced. The PP&C organization must still be involved in some or all of the following: (1) providing schedules to the final workstations that distribute workload throughout the production system; (2) working closely with suppliers to reduce lead times and manage their capacity more effectively in order to provide just-in-time delivery of parts more inexpensively; (3) assuming a greater role in interfacing between marketing, design, sales, and the shop floor—particularly in smoothing the introduction of new products onto the shop floor; and (4) providing support in establishing the pull system, at least during early stages of implementation. These responsibilities coupled with the simpler line arrangements in agile production mean that the emphasis for the PP&C organization shifts away from controlling execution on the shop floor and releasing orders to vendors. Instead, PP&C spends more time managing capacity despite shifts in product mix and establishing better coordination mechanisms that link suppliers to the production system.

Quality Assurance

The role of QA in the conventional factory is to make sure that quality standards set by the product design department are rigidly enforced. To reduce the cost of inspection, quality engineers typically rely on sampling of incoming parts, on-line components, and finished assemblies. In the best of circumstances, sampling is clearly a reactionary measure. When defects are detected by sampling, solutions are normally costly. For example, it may be necessary to rework or scrap large quantities of parts. In the case of poor-quality supplied parts, it may be necessary to return them or to use them in some unintended fashion if lead times prohibit returning.

In the agile production system, the goal is perfect product quality every time. To accomplish this, product quality becomes the responsibility of the workers producing the parts, product engineers, and process engineers. Consequently, the role of QA shifts from the role of police officer to one of facilitating improvement. For example, QA may focus on training and educating workers in quality concepts such as statistical process control, self-checks, and problem identification and analysis. The QA organization may also work with vendors to help raise the quality of vendor products. For example, incoming inspections of supplier parts can be eliminated by developing and maintaining close relationships with a few quality suppliers whose products can be trusted. In addition, QA may perform the more complicated technical inspections, chemical analyses, X-ray analyses, destructive tests, process capability studies, and so forth.

The real responsibility for quality, however, rests with shop floor workers. To catch quality problems quickly, workers check one another's work at each successive workstation. Data are collected and measurements taken at each workstation to track quality over time and to identify problems as they arise. When a quality problem is found, production is temporarily halted. Shop floor workers, with the support of production engineers when needed, determine the source of the problem and how it can be corrected. Once the problem is solved, the line is restarted. Modifications to work procedures or tooling may be made at the source workstation to avoid recurrences of the same problem.

This sophisticated system of shop floor quality feedback and control is greatly enhanced by the line layout and product orientation. The product-orientation ensures close proximity of workstations to one another and efficient communication of quality concerns as they arise. The smaller batch sizes and reduced changeover times also encourage faster feedback on quality. Finally, the manufacturing philosophy of the agile production system encourages extremely high standards of housekeeping as well as the discipline needed to sustain high quality.

In the agile production system, each member is an integral part of the production activity. Each person or work team services someone and is served by someone. The internal chain of customers links the work teams together into a single unit that strives for continuous improvement. When large-scale or rapid improvement is needed, task teams are appointed to make changes happen. Task teams may need training in the tools and practices needed to redesign the work environment—training that can be provided by the QA department.

Within the agile production system, the significance of and attention to quality is greatly increased. Quality is seen as a key to customer satisfaction, achieved through continuous work process and product improvement. The role of QA in this environment becomes that of facilitating continuous improvement through facilitation, training, and equipping teams with the tools, skills, and resources necessary for streamlined operations.

Purchasing

Purchasing in the conventional production system involves locating suppliers, arranging and negotiating agreements, and following up on orders to assess the quality of the raw materials and parts that are acquired. Purchase orders are based on production needs and are generally large to ensure that bulk discounts are applied. Incoming lots received from the supplier must be inspected and compared to incoming invoices to detect variances. When problems are detected, personnel in design or quality engineering are called in to make judgments or change the specifications.

Supplier selection in conventional organizations is almost always guided by management policies requiring the selection of the lowest bidder. Consequently, buyers must maintain information on prices, new products, and materials for a large number of suppliers. In this system, buyers tend to be overwhelmed by the sheer amount of product-related information and with the number of supplier relationships. The buyer is forced into attending closely to cost, adhering to specifications, and negotiating the contracts. Questions or comments from suppliers concerning materials or production processes are often ignored, taken as criticism, or passed on to the appropriate engineer, resulting in delays in establishing contracts.

Unfortunately, parts purchased on the basis of lowest bid may be of inferior quality and may directly cause major production problems and delays. A single supplied part that is out of tolerance on the line will either shut down production or cause assembly of defective products. In addition, the lengthy storage times for raw materials and parts from multiple suppliers make tracking quality problems back to suppliers difficult or impossible.

Purchasing in the agile factory differs radically from that in the conventional system. Figure 1.3 lists some of the key characteristics of purchasing in the new system. The emphasis is on small lot sizes with frequent (just-in-time) deliveries and on establishing quality relationships with a limited number of suppliers. In this system, buyers devote most of their time to finding good suppliers, arranging flexible single-source agreements, and maintaining quality relationships with these suppliers. Single-source relationships require more time per relationship, but, overall, are much easier to manage than dealing with many competing suppliers. For example, paperwork needed for a given part can be greatly reduced, contracting and ordering is greatly simplified, and contracts specifying quantities to be delivered over several months

---

Characteristics

  Small lot sizes with frequent deliveries

  Reduced number of suppliers

  Selection of suppliers based on quality, not cost

  Close, cooperative, long-term relationships with suppliers

  Involvement of suppliers in purchasing and production activities

  Redesign of delivery system to ensure mixed-load, on-time delivery

  Redesign of receiving area to eliminate inspection and improve efficiency of parts handling

  Stabilization of parts scheduling, including potential kanban system with suppliers to draw parts into system

  Standardization of shipping containers

Figure 1.3. Purchasing in the agile manufacturing system.

can be established. These long-term contracts encourage the supplier to essentially become an extension of the agile production system. Teams of production personnel may, in fact, visit the suppliers to examine production processes and quality orientation and make suggestions that will improve overall production performance. Close supplier–customer linkages allow orders to be triggered directly from the shop floor personnel, who send "production cards" requesting needed parts and materials. These production cards are returned with the supplied materials and are used as invoices for receiving.

Parts are transported to the agile factory more frequently and in smaller quantities than in conventional manufacturing. As a result, suppliers may be chosen who are in close proximity to the factory. To improve efficiency, a single delivery truck may pick up small loads from several vendors rather than relying on independent deliveries from each supplier. Ideally, parts are delivered directly to the line, avoiding warehouses and other forms of storage.

Clearly, a key role of buyers in the agile production system is the selection of suppliers. Rather than relying on the lowest bids, the buyers focus on product quality and delivery times. Other factors are incorporated into the supplier selection process as well, including supplier experience, management philosophy, work processes, location, financial stability, labor relations, willingness to improve, and contract management. It should be stressed that the buyer in this new environment does not necessarily make a decision alone, but often works with cross-functional teams to make these decisions. These teams may include representatives from engineering and production who will be directly affected by the decisions.

Overall, the buyer's role in the agile factory is significantly more complex and demanding than that of the buyer in a conventional factory. The buyer in an agile factory must be knowledgeable in a variety of technical areas and possess strong interpersonal skills. Broad knowledge of agile manufacturing practices is required so that materials and parts are ordered according to the needs of the production system rather than the needs of the financial system. Knowledge of statistical methods is needed to assess supplier quality and performance. Buyers may even need to manage organizational change, as suppliers are encouraged to adopt continuous work process and quality improvement and agile production methods. Excellent interpersonal skills are required to maintain close relationships with suppliers while meeting the needs of production.

Maintenance

Maintenance, or plant engineering, is responsible for machine tool and equipment repair; in-plant construction; and repair of other mechanical, hydraulic, or electrical problems that may or may not be directly related to production. Conventional maintenance organizations are reactive. That is, maintenance mechanics are typically dispatched to the site of machine breakdowns after they have occurred. Because of their busy schedules, mechanics often do not have time to investigate the cause of problems or to provide machine operators with advice on maintaining and monitoring the equipment to prevent future failures.

The relationship between production and maintenance causes additional complications. First, because preventive maintenance and equipment repair are maintenance functions, machine operators do not see preventive maintenance as a personal concern and may ignore dirty equipment or minor maintenance needs. The incentives to meet production goals and the punishments for failing to meet these goals make operators far more interested in production than in taking care of their equipment. Further, operators in the conventional system generally lack the training to make minor adjustments and may even lack the training needed to operate their equipment properly, causing early and frequent equipment breakdown. Thus, operators may not understand the importance of preventive maintenance and may feel reluctant to ask for help from maintenance mechanics. These factors result in poorly maintained equipment, frequent breakdowns, missed production goals, and an ongoing adversarial relationship between maintenance and production workers.

In agile manufacturing, the goal of maintenance is the complete elimination of all equipment breakdowns and problems. To accomplish this goal, equipment operators take on many of the routine maintenance and preventive maintenance activities normally done by maintenance personnel. Maintenance crews are expected to perform the more difficult preventive maintenance activities and to provide training and education to operators regarding the care and operation of the equipment. Operators have responsibility for not only using the equipment properly, but also for the mechanical performance of the equipment. In part, the philosophy behind this expanded role is that if operators are responsible for equipment performance, they will be more sensitive to the care and maintenance of the equipment. Figure 1.4 includes some of the expanded operator tasks. Note that these tasks include participating in training dealing with equipment operation and functioning, conducting routine preventive maintenance, performing daily cleaning and inspection, and working closely with maintenance personnel during repairs to get a better understanding of equipment functioning.

Assigning routine maintenance responsibilities to operators reflects a considerable shift in maintenance philosophy. The close attention to the care of equipment by operators is critical to minimizing unexpected breakdowns and the resulting loss in production. Such maintenance also helps prevent "process drift," in which product quality gradually deteriorates with declining equipment performance;

Operator tasks

Performing housekeeping to maintain a clean and orderly workplace

Looking for potential sources of dust, dirt, and other contaminants that might interfere with equipment functioning

Cleaning and inspecting equipment daily

Conducting routine preventive maintenance, such as lubricating, tightening loose bolts, and so on

Monitoring equipment performance and, when a problem arises, taking steps to ensure the problem is corrected

Participating in training to learn more about equipment operation and functioning

Working with maintenance personnel during regular preventive maintenance to expand knowledge of equipment

Maintaining records of equipment performance

Figure 1.4. Maintenance tasks assumed by operators in agile production system.

increases equipment service life; and reduces accidents due to equipment malfunctioning.

Although operators in the agile factory do everything possible to avoid breakdowns, some will still occur. The role of maintenance is then one of emergency response and support. Since maintenance has no routine preventive maintenance duties and a far lighter load of emergencies, it is able to respond promptly. Having responded to the immediate need, maintenance and production engineering personnel work cooperatively to identify equipment modifications that can prevent the problem from reoccurring.

Overview of Production Support

In conventional manufacturing, the production and the production support organizations pursue their own goals and are often involved in adversarial relationships with each other. The resulting suboptimal performance means that the organization as a whole cannot hope to be globally competitive. However, in the agile production system, each production support function is aligned with the production system. Each of these functions tends to get smaller as key tasks are taken over by the production system. The focus on resolving problems and "fire fighting" declines as more attention is given to quality and to problem prevention. While the production system moves to a larger and more central role in the organization, each of the production support organizations adopts a strong customer service orientation. The result is the task-aligned organization, in which overall performance is optimized as all functions pull together for the common goal.

Business Operations

The third major organizational grouping, business operations (see Figure 1.2), provides a link between the organization and its external business environment. Like production support, business operations take on a considerably different role under the agile production system. Engineering works with marketing and production to speed up the product development cycle. Marketing and sales take advantage of agility in manufacturing to build a new sales strategy based on quality products and rapid response to customers' needs. Human resources attracts and trains individuals capable of performing within an agile environment. Finance and accounting measure agility and encourage investment toward a future agile manufacturing vision. Each of these functions is aligned in support of the production system to provide critical services. When these functions are properly aligned in support of the production system, the manufacturing organization as a whole is able to compete successfully within the new global business environment. The following sections provide more detail about the role of each of the business operations functions in the agile production system.

Engineering

Engineers are widely needed throughout manufacturing organizations, with the exact needs depending on the type of product being produced. Most conventional organizations rely on both product and manufacturing engineering staffs. Product engineers are responsible for designing a product that meets customer requirements, while manufacturing engineers are responsible for manufacturing the product. The design process usually occurs in three stages. First, product engineers work with marketing personnel to define the customer requirements. Second, a model or prototype is designed that will achieve the requirements established in the first stage. Third, product engineers work with manufacturing engineers to finalize a design that can be manufactured and assembled economically.

This approach to engineering was developed years ago, when products were technologically much simpler, model changes were very infrequent, and the emphasis was on creating product volume. Both product design and product manufacturing were much simpler processes at that time. However, in the current manufacturing environment, product changes are quite frequent. New technology, new materials, and new designs are constantly replacing the existing approaches. Not only do products change, but product features must often be tailored to meet customers' needs. Conventional engineering approaches are simply inadequate to meet these demands for change and response.

In the agile production system, however, product and manufacturing engineering are conducted concurrently. Product and manufacturing engineers work together in teams, along with representation from other functions (for example, marketing, purchasing, and quality) to design the product. Such teaming improves communication among participants and helps avoid costly design changes in later stages of the product development cycle. The project managers assigned to lead these product development teams keep designs goal-focused and time lines on schedule.

A primary goal of design engineering under the new paradigm is to design products that both meet customer requirements and are exceptionally easy to manufacture. Ease of manufacture is increased by designing in common parts across several different products. Using readily available parts not only saves money and time, but minimizes the overall number of different parts to be used. Also important is redesigning existing parts to simplify assembly.

The design teams also use computer-aided design (CAD) systems to produce initial images of the product and use other engineering software to model and evaluate the manufacturability of the designs, to estimate the product/process interaction, and to suggest better materials or processes for manufacturing the product design. These engineering software systems allow new customer requirements or design changes to be quickly incorporated into the manufacturing process with much less need for costly and time-consuming model or prototype construction.

Manufacturing engineers in the new environment retain an important role in new equipment design, specification, and implementation. However, increased emphasis is placed on obtaining or developing equipment that is easy to reconfigure for handling mixed-model production. This equipment is built to be easily moved, as the layout of the shop floor is frequently revised to meet new requirements, while maintaining smooth flow. Particular attention is paid to finding creative ways to eliminate the need for traditionally centralized process equipment like heat treatment, painting, plating, and cleaning in order to improve layout flexibility and simplify shop floor material handling.

Manufacturing engineers in the agile production system support continuous improvement efforts by helping shop floor workers with the analysis and solution of production problems. They can also help to identify, develop, and/or procure training needed by shop floor workers to improve their skills. Note, however, that the primary responsibility for continuous improvement remains with the production workers. The workers themselves are expected to monitor equipment functioning and production performance, notice problems as they arise, and suggest ways of improving the process. When problems arise that are beyond the ability of production workers to solve on their own, then engineers are called in for support and consultation. The support role for engineering is much different than in the conventional manufacturing system, in which the engineers may assume primary responsibility for dealing with production problems. Without a firm customer service orientation, the engineering staff may be tempted to bring new technologies or processes into the line just because (the engineers) think they are useful. In the agile production system, engineers are definitely encouraged to propose changes and improvements, but they must sell their ideas to the line and line management.

Marketing

In manufacturing organizations, marketing is primarily interested with defining and selling products. Specific responsibilities include pricing, advertising, sales forecasting, and conducting market surveys. In conventional organizations, the sales forecast is perhaps the most important responsibility of marketing. The sales forecast helps to coordinate the various business departments around production goals. For example, forecasting information is used by production planning to determine capacity and material requirements. Product engineering and product research and development (R&D) may be urged to speed up or to abandon the design and development of certain product lines, depending on predicted product forecasts. Finance relies on sales forecasts to establish future budget. Human resources keys its recruiting and hiring activities to the forecasts. Marketing uses its own forecasts to develop and target sales strategies. Clearly sales forecasts have broad organizational impact.

In the agile production system, the focus of marketing is on capitalizing on the benefits of agility, turning these benefits into a distinct competitive advantage. These benefits include low prices, reduced response times, high quality, the ability to provide customized features, excellent customer service, and reduced delivery times. Marketing must be able to communicate these advantages to customers, increase sales by capitalizing on the competitive advantages, and ensure rapid response time to changes in customer needs.

While marketing personnel in the conventional factory may spend a great deal of time writing customer requirement documents, in the agile production system, these personnel are actively involved with the production process. Marketing staff have direct and frequent contact with customers and quickly provide information from these contacts to other internal organizations. For example, marketing facilitates the development of new products by providing information about customer needs to product engineering, to product design teams, and to production personnel. Marketing is also a key source of real-time feedback on customer satisfaction and on performance of the production system. To provide this vital information, however, marketing personnel actively seek information on customer needs and customer satisfaction with products and services. That is, marketing personnel *listen* as well as market products. They also work closely with QA to follow up on issues of customer satisfaction and to communicate these issues throughout the production system. In fulfilling these roles, marketing becomes an important communication link between the customer and a number of functions within the agile organization.

Human Resources

The primary purpose of the human resource (HR) organization should be to provide a highly-motivated, competent, and well-trained workforce throughout the organization. Specific services would include recruiting competent workers, maintaining standards of behavior, rewarding good performance, correcting deviant behavior, communicating with labor unions, monitoring equal opportunity employment compliance, and maintaining classification and compensation systems.

Unfortunately, in many conventional organizations, HR tends to be more focused on protecting the organization from its employees and to concentrate on administering numerous rules and policies. Following the rules seems to become virtually an end in itself. As a result, the HR function tends to become very bureaucratic and authoritarian, completely losing sight of its service role. The HR function may exert considerable pressure on other parts of the organization to conform to established policies, rather than trying to provide optimum services to these functions. That is, HR may suboptimize overall organizational performance in its attempt to optimize its own performance.

These actions adversely affect manufacturing performance. For example, rigid job classification policies restrict the ability of workers to rotate between tasks, to take on preventive maintenance and quality responsibilities, or to make production decisions. Compensation policies limit the ability of managers to reward team performance, rather than individual performance. Job promotion policies inhibit worker motivation and loyalty.

In contrast, HR within the agile production system is highly customer-oriented and supportive of the innovations being introduced into the manufacturing process. Workers and crews are rewarded for their new ideas and increased efforts. Performance appraisal systems are implemented that reward and encourage performance, rather than humiliate and anger employees. The HR function serves as a leader in efforts to streamline procedures, lower costs, and reduce red tape. Table 1.2 provides some of the key characteristics of the HR organization needed for agile manufacturing.

In the new paradigm, HR becomes customer-oriented, with its customers being the other functions within the organization—especially the production system. HR works with its customers to implement policies supporting job rotation, employee recognition, and self-managed work teams. Workers are rewarded on the basis of skill and knowledge acquisition as well as company performance and success. In addition, HR becomes much more of a facilitator for human resource development. Recruiting and interviewing may be

Table 1.2. Human resources in agile production system.

Characteristic	Description
Customer orientation	Provide high-quality service. Production organization becomes a key target group.
Recruiting	Involve line managers and workers. Interpersonal and team skills are key criteria.
Rewards	Adopt pay-for-knowledge, profit sharing, and incentive pay. Reward teamwork.
Performance appraisal	Participative goal setting and ongoing feedback.
Training	Cross-train in production. Process improvement training driven by problems in work environment.
Work policies	Adopt policies supporting production-initiated improvement.
Classification	Take into account continuous improvement and employee empowerment.
Employment security	Retain skilled workers. Retraining and redeployment are key.
Labor negotiation	Identify union requirements as legitimate.

arranged and scheduled by HR, but they are conducted by line managers and workers. While a bonus and recognition program may be initiated by HR, work teams themselves acknowledge outstanding individual and team performance as well as determine training and education needs. HR works with each organizational unit to identify personnel requirements and implement policies consistent with each. In this way, the HR organization aligns itself with production.

Finance and Accounting

The role of finance and accounting is to oversee the administration and control of the company's assets. This includes functions such as internal capital financing, external investment analysis, budgeting, financial reporting, cost accounting, and data processing. The major requirement for performing these functions is the development and maintenance of a management control system.

The problem with many existing management control systems is that they are based on concepts developed nearly a century ago to support labor-intensive methods of mass production. For example, many existing accounting systems assign overhead charges on the basis of direct labor charges. This approach was satisfactory when direct labor costs formed the large majority of production costs (often 80 percent or more). However, in recent years the proportion of total production costs due to direct labor charges has plummeted to 20 percent or less, while overhead costs, such as product and manufacturing engineering, have greatly increased. Nonetheless, conventional management control systems rely on outdated performance measures such as the ratio of indirect to direct costs. Further, conventional cost-accounting systems fail to accurately determine product costs (Hall, Johnson, and Turney 1991). In short, conventional cost-accounting systems fail to provide the cost and production information needed by management to make informed production decisions.

Current accounting systems are also vulnerable to manipulation. By minimizing capital investments in facilities, equipment, and research and development, management can lower overhead expenditures. The conventional performance measures—which focus on return on assets—are thus made to appear artificially good in the short-term.

In the long-term, however, these practices are extremely harmful to the company's competitiveness because they stifle product and process improvement. For example, by not replacing worn-out facilities and equipment, increasing demands are placed on workers to maintain production levels in spite of these restrictions. Consequently, workers are placed at a disadvantage in comparison to the workers of competitors who have better facilities and equipment.

Production line managers can manipulate conventional control systems in other ways as well. Reports on scrap rates can be intentionally inflated to hide the creation of a reserve of parts to serve as a backup in case of delivery problems. Budgets for employee training and equipment maintenance can be carefully hoarded to compensate for cost overruns in materials and labor. As a last resort, managers or supervisors may charge materials to improper charge accounts to preserve the appearance of sound fiscal performance.

In the agile production system, measures of organizational performance are developed that support organizational effectiveness and continuous improvement. These measures focus on quality, inventory, productivity, innovation, and status of the workforce. Table 1.3 provides examples of measures that might be used in an agile production system.

Table 1.3. Examples of performance measures in agile production system.

Type of measure	Examples
Quality	Defects, per million rework, defect rate reported by customers, cost of quality
Inventory	Average batch size, work-in-process, inventory
Productivity	Units produced, materials consumed, energy used, labor hours needed, number of equipment breakdowns
Innovation	Number of new products, number of modifications, improved performance characteristics, rate of quality improvement
Workforce status	Attitude, morale, improvement in skill levels, turnover, absenteeism

The purpose of these measures is to provide the production system with accurate and useful feedback on its performance. Quality measures, used in conjunction with statistical process control, help to assess the stability of the manufacturing process and to determine when intervention is necessary. Inventory measures help focus on the reduction of key agile manufacturing goals in the area of reduced work-in-process and inventory. Productivity measures have a production and work process orientation, rather than the dollar or financial orientation of conventional performance measurement systems. In the agile production system, when workers or management strive to look good on performance measures, they do so only by increasing actual performance, not by manipulating inappropriate measures.

In the agile production system, the finance and accounting functions make no attempt to control production activities. The proper role of these functions is to gather, analyze, and communicate organizational performance information in support of production activities. Finance and accounting, like the other support functions discussed in this chapter, are task aligned and promote overall organizational effectiveness.

Conclusion

The overall vision of the agile production system involves complete task alignment of all support functions, centered around the production system, with all functions operating from a philosophy of agile (fast and flexible) work processes and continuous work process improvement. Machines are made flexible through the reduction of setup and changeover times. Plant layouts are based on a product orientation, capable of rapid reconfiguration in response to new customer demands. Production control, quality control, purchasing, and preventive maintenance are closely integrated into the production environment. Marketing, engineering, human resources, and finance operate in support of production effectiveness and improvement and increased competitiveness of products.

Agile manufacturing is characterized by a strong customer service and team orientation. The production system functions as a team dedicated to serving its internal and external customers. The production support and business operations functions subordinate their own interests to those of the production system and to the organization as a whole. The result of achieving task alignment and agile manufacturing is high product quality, the ability to respond rapidly to changing customer demand, and greatly improved organizational effectiveness.

References

Beer, M., R. A. Eisenstat, and B. Spector. 1990. Why change programs don't produce change. *Harvard Business Review*, November–December, 158–66.

Black, J. T. 1991. *The design of the factory with a future*. New York: McGraw-Hill.

Hall, R. W., H. T. Johnson, and P. B. Turney. 1991. *Measuring up: Charting pathways to manufacturing excellence*. Homewood, Ill.: Business One Irwin.

Schonberger, R. J. 1990. *Building a chain of customers*. New York: Free Press.

CHAPTER 2

Managing Systemwide Change
Joseph C. Montgomery

Based on results, creating an agile manufacturing organization is a significant task indeed. As with any large-scale change effort, numerous issues must be addressed in order to achieve a successful outcome. Failure to address these complex issues helps explain why, of the many organizations that have embarked on change programs over the last few years, so many organizations have failed. For example, Majchrzak (1988) indicated that 50 percent to 75 percent of U.S. firms experience failure when trying to implement some form of advanced manufacturing. Nadler and Tushman (1989, 100–19) indicated that fewer than one change program in 10 was successful when the change was attempted in reaction to environmental conditions rather than in anticipation of them. Similarly, Schaffer and Thomson (1992) found that 63 percent of the electronics firms surveyed who had initiated a total quality management (TQM) program had failed to achieve even a 10 percent reduction in defects. Scott-Morgan (1994) conducted a survey of 350 major companies in the United States and found that only 17 percent were satisfied with the results of their major change efforts—40 percent were positively dissatisfied, and 70 percent complained of unanticipated problems and side effects. Other writers (for example, Beer 1987; Beer, Eisenstat, and Spector 1990; Chew, Leonard-Barton, and

Bobn 1991; Kilman 1989, 200–228; Kilman 1991; Lawler 1991) have also stressed the tendency for change efforts to fail.

This chapter examines the reasons for the success or failure of change programs and presents a model of change that learns from these mistakes and can be used to help avoid the pitfalls experienced by others. It should be noted that several of the examples provided are based on large organizations, rather than on the midsized or smaller organizations on which this book focuses. While in part these examples were chosen because of the limited literature available on midsized organizations, the main reason was that the basic principles of system change are the same regardless of the size of the organization. Further, as will be discussed later, these principles of change hold true regardless of level within the organization—whether one is working at the corporate, division, department, team, or individual level. On the other hand, while change principles remain the same, the sheer complexity of issues that arise and the real difficulty encountered in managing change certainly do increase with organizational size as well as with organizational level. By using larger organizational examples, we present the unchanging principles to managing system change while including the worst-case scenario for complexity and difficulty.

Reasons for Success or Failure of Change Efforts

A number of explanations have been proposed as to why some change efforts fail while others succeed. Reasons for failure have included reliance on the "quick fix" (Kilman 1991), the "fallacy of programmatic change" (Beer, Eisenstat, and Spector 1990), use of activity-centered rather than results-oriented change (Schaffer and Thomson 1992), the difficulty in creating change due to system complexity (Senge 1990), and the use of inappropriate change strategies (Lawler 1991).

The Quick Fix—Failure to Take a Systems Approach

Ralph Kilman, in *Managing Beyond the Quick Fix* (1991), has argued that companies become more difficult to change as they grow older and, like humans, suffer from "hardening of the arteries." These companies become burdened with an accumulation of policies, strategies, and a culture that may have been successful in the past but is not appropriate for the present or future. Rather than attacking these issues broadly, management is tempted to adopt a single program that poses some promise of success. For example, management might adopt a new strategic planning system, develop a new format for rewards and recognition, or change the pay structure. These single-approach solutions, referred to as quick fixes or "band-aids," simply treat symptoms and ignore large systemwide problems. Kilman advocated a systems theory–based change management program to accomplish change in which the entire organization is considered.

The Fallacy of Programmatic Change

In an article entitled "Why Change Programs Don't Produce Change," Beer, Eisenstat, and Spector (1990) looked at success and failure of change from a different perspective. According to these writers, companies that have succeeded at change have accomplished the following:

- Reduced the role of management authority, rules and procedures, and jobs with narrow scope
- Created teams and encouraged teamwork
- Shared information
- Pushed responsibility and accountability down into the hierarchy
- Moved from a bureaucratic organizational model to a task-aligned model (similar to that described in chapter 1).

Unfortunately, most companies have not been able to accomplish these results. Instead, many companies have relied on *programmatic change*, defined as organization-wide training programs or other broad programs of change that do not have specific improvement targets. Beer, Eisenstat, and Spector (1990) labeled the failure of such programs as "the fallacy of programmatic change," citing their four-year study of six large corporations that documented the negligible impact of company-wide change programs. These programmatic efforts included strategic planning, corporate culture programs, pay-for-performance systems, quality circles, and large-scale training efforts. Programs were seen as likely to fail if driven primarily by top management or if sponsored by staff groups such as the human resources department.

Further, as one program failed in these organizations, it tended to be replaced by another, with the result that "wave after wave of programs rolled over the landscape with little positive impact" (p. 159), just as the sequence of quick fixes described by Kilman had little impact. The rapid progression of unsuccessful programs served to undermine the credibility of subsequent programs, further decreasing the likelihood of success.

Beer, Eisenstat, and Spector (1990) felt that these programs had failed because of their inability to generate (1) coordination and teamwork, (2) commitment, and (3) new work-related competencies. For example, a well-written strategic plan would be unlikely to elicit commitment among organization members, particularly if it was written only by management or by a planning group. Changes in the organizational structure might change the organizational chart but have no impact on the attitudes and skills needed by workers. Similarly, the skills taught in training programs would tend to do nothing to improve coordination and, in fact, might cause resentment when the workplace was not receptive to using the skills. The authors state, "Because they are designed to cover everyone and everything, programs end up covering nobody and nothing particularly well. They are so general and standardized that they don't speak to the day-to-day realities of particular units" (p. 161).

Activity-Centered Versus Results-Centered Change Programs

Other writers have not criticized programmatic efforts per se, but instead have analyzed the failure or success of these programs in terms of their focus and orientation. Schaffer and Thomson (1992) argued that activity-centered programs, which often advocate a particular philosophy or management style, contribute little or nothing to bottom-line company performance. On the other hand, results-oriented programs that focus on achieving specific, measurable results within a specified time frame are often extremely successful.

Schaffer and Thomson compared the impact of activity-centered programs on organizational performance to the impact "a ceremonial rain dance has on the weather" (p. 80). The activities (which may be pursued with a great deal of zeal) might be TQM, continuous

improvement, empowerment, employee involvement, customer satisfaction, or any other change program. The erroneous assumption is
that once the prescribed analyses, training, and meetings are completed, there will be some bottom-line improvement in quality, sales,
and inventory. That is, if the programs are religiously followed, results
will "take care of themselves." For example, a financial institution initiated a TQM program, training hundreds of employees and communicating the program to thousands of others. In the subsequent
evaluation of the effort, results of the program were couched entirely in
terms of training completed, teams created, and morale—there were no
real bottom-line performance results to report.

Many other companies have also expended great sums of money
on programs that have come to naught. Schaffer and Thomson pointed
out that managers are often under extreme pressure to solve problems
and to keep up with competitors. The pressure encourages them to
adopt almost any reasonable-sounding approach. Because many of the
programs, such as TQM, have good reputations and are espoused by
numerous professional societies, academics, research, and well-known
consultants, it is perhaps only natural to place faith in them. However,
any good approach to improving organizational performance must be
linked with an effective change management strategy. The managers
who fail to get positive results are often those who have overlooked the
implementation issue. (Note that the latter half of this chapter is
devoted to describing such a change management process.)

Schaffer and Thomson argued that activity-centered programs are
likely to be disappointments for six reasons.

1. *The programs are not linked with specific results.* The training
may identify general principles, but may not specify the desired end
state. After team training in agile manufacturing, for example, each
work team should know exactly what is expected in terms of work-
process analysis and improvement and must be linked with other
resources who will guide, coordinate, and support their efforts. The
actions of each team are part of a larger vision of change. However, in
activity-centered training, workers may go to training as individuals,
rather than as a team, and may have no further contacts with anyone.
Thus, no mechanism is created for organizing and aligning team

efforts. Further, different organizational units may go off in radically different directions, with no coordination or agreement on proceeding with improvement efforts. The organizational infrastructure needed for bottom-line results has not been developed to complement the training.

2. *The programs are too large-scale and diffuse.* Large numbers of employees may be trained, but their subsequent actions are not linked by any vision or sense of direction from management. Individuals are expected to somehow implement complex programs.

3. Results *is a four-letter word.* Managers have been frequently attacked for their bottom-line, short-term orientation (for example, Peters and Waterman 1982). As a result, a bottom-line orientation has become less fashionable. In their desire not to be accused of these failings, managers may overcompensate and fail to demand results.

4. *Delusional measurements are used.* Activity-centered programs tend to measure success by the number of persons trained or groups formed. These are inappropriate criteria of success.

5. *The programs are driven by staff or consultants.* Assigning change programs to these groups serves only to relieve busy managers of immediate responsibility. Real success comes only when driven by line or operational managers who know the work and are accountable for results.

6. *The programs are biased to orthodoxy, not empiricism.* Programs, especially off-the-shelf versions, are guided by how things *should* be done rather than by using actual projects and test cases to drive organizational learning.

In short, activity-centered programs are founded on the belief that a change in attitude will lead to a change in behavior, and that changes in individual behavior, repeated across the organization, will lead to organizational change. The authors argued that the reverse is more accurate: that behavior is influenced by placing people in new organizational roles that impose new requirements on behaviors.

Schaffer and Thomson strongly recommended using the results-oriented approach to change. As an example, they cited the case of a bank that needed to become more competitive to survive. The heads of each of 20 organizational units were each told to select one or two customer improvement goals that could be implemented quickly. While several workshops were held and consultants were used, a clear focus was kept on improvements that would help accomplish the goals. Within a year, several million dollars of savings were obtained as all 20 units achieved their goals. (In a similar time period, an activity-centered program might just be finishing designing and implementing training.) The lessons learned from this example included the following:

1. *Introduce managerial and process innovations only as they are needed.* That is, goals should be prioritized and employees should receive training on a just-in-time basis.

2. *Rely on empirical testing.* Week-by-week goals allow data to be collected quickly and frequently. Improvements that fail to work out can be dropped and those that are effective used more broadly.

3. *Use frequent reinforcement to energize improvements.* Short-term projects provide immediate success feedback that is highly motivating to both employees and managers, as compared with the long-term, often fruitless efforts of activity-centered programs.

4. *Create continuous organizational learning.* By using smaller, short-term projects, management and workers develop a base of experience in change that can be applied organization-wide.

In order to implement a results-oriented change effort, the authors suggested that top management first request unit managers to identify a few areas for improvement and to set ambitious but short-term goals for accomplishing the improvements. The unit managers then are responsible for seeing that the goals are achieved. They conduct whatever training is needed, establish teams that are needed to accomplish the goals, collect data on progress, and report regularly to top management. As successes are built up across the organization, the change efforts are then institutionalized through policy, rewards and incentives, and organizational structures (for example, an improvement office). Top management continues to participate by developing a clear vision of where the changes should be leading the organization.

Difficulties of Change Due to System Complexity

Peter Senge in *The Fifth Discipline* (1990) focused on the importance of systems thinking to avoid a number of possible problems associated with change. He argued that the built-in lag time between management actions and feedback about changes in organizational performance is so long that it is often difficult to distinguish effective from ineffective actions. The following are some of the ineffective strategies of systems change that he identified (note that he includes broad sociopolitical as well as organizational examples).

1. *Shifting the problem to a different part of the system.* Here, attempts at solving an organizational problem simply shift the difficulty elsewhere. For example, relying on rebates to encourage sales may lead to a temporary surge in sales, then a lasting slump as customers develop a resistance to paying the original price. Another example would be an organization hiring staff members to deal with specific financial and administrative issues. While these problems may be solved, eventually the organization becomes loaded down with overhead personnel and is no longer competitive. Perhaps the clearest example is that of arresting drug dealers in one particular neighborhood. Drug sales are reduced at that location, but shortly after, sales pick up sharply elsewhere as the dealers set up shop in a new location. The moral is that solutions that merely shift problems to different parts of the system are no solution at all.

2. *The problem of compensating feedback,* or "the harder you push, the harder the system pushes back." This problem of systems change is illustrated by the attempt in the 1960s to solve housing problems by spending a great deal of money building low-cost housing in central cities and providing family financial support. The result a decade later was even worse poverty, crime, and unemployment as an unanticipated flood of low-income people migrated to cities with the best programs, overwhelming these programs with applicants. As an organizational example, consider a company redoubling marketing efforts on an older product that is showing a decline in sales. While sales may increase in the short-term, the fact that funds were diverted from the rest of the company results in reduced development of new products. Long-term sales therefore drop dramatically, placing the company in

severe financial straits. The underlying issue, according to Senge, was that in pushing harder to solve perceived problems in a system, the time delay between actions and feedback increases the likelihood that we will lose sight of how we are contributing to the problem ourselves.

3. *Bad solutions often work well at first,* or "behavior grows better before it grows worse." For example, suppose a manager wants to move up in the organization and engages in political maneuvers in order to look good. While there may be some short-term success, the likelihood is long-term failure as managers and peers alike realize the manipulative tactics that are being used. As a second example of this problem, consider the strategy of a manager coming down hard, even to the extent of being abusive, on workers who are perceived as not being productive enough. While there may be short-term increases in performance, the likelihood is of long-term decreases in productivity because not only were the real problems not addressed, but resentment and hostility have been increased. The only real solution to problems in a system are those that work well in the long-run.

4. *Familiar solutions do not work when fundamental problems persist,* or "the easy way usually leads back in." Here, Senge noted that the solutions we are familiar with tend to be the ones we apply when problems arise. However, these solutions may or may not be what is needed. For example, a business may focus on cost-cutting measures in an attempt to remain competitive rather than examining and improving the basic production processes. In the context of politics in the United States, to improve the U.S. economy, Democrats tend to initiate new programs, and Republicans tend to reduce taxes to businesses. Neither party tends to examine the highly complex problems affecting competitiveness.

5. *Shifting the burden to an intervenor,* or "the cure can be worse than the disease." Here Senge refers to situations where the easy cure proves to be not only ineffective, but also addictive and dangerous. An example of this type of problem would be, according to Senge, applying ill-conceived government interventions to schools, housing, education, and so forth. These solutions are insidious because gradually more and more of the "solution" is needed as dependency and reduced capabilities are fostered. A straightforward example would be relying

on a calculator to do all of our mathematics, resulting in less understanding and capability in mathematics. An organizational case might involve a CEO bringing in an outside consultant to help with modernization efforts, failing to thoroughly learn and understand the consultant's approach, and finding a growing dependency on the consultant to solve difficult problems. The underlying theme of this problem is that, over time, the intervenor's strength and size grows, causing worse problems than the problems it was designed to cure.

6. *Faster is slower,* or "the optimal rate of change is far less than the fastest possible rate." The issue here is that when growth or change occurs too quickly, the system compensates by slowing down. Quick fixes simply do not work. Examples of this type of thinking would be to try to cure productivity problems by just bringing in TQM, management by objectives (MBO), participative management, or new equipment in a hurry and expecting the problem to be solved. Similarly, many managers think they can change the entire company overnight by writing a new mission statement and a nice strategic plan. These approaches do not work. Real system change will take time.

7. *Cause and effect are not closely related in time and space.* This represents a key problem in dealing with systems issues. Effects here refer to obvious symptoms of systemic problems (such as low productivity, high turnover, unemployment, drug abuse, starvation) responsible for the effects and which, if recognized, could lead to changes and lasting improvements. The basic problem is that we expect simple, direct linkages between causes and effects when the reverse may be true. We know from cognitive psychology that human beings possess limited information processing capabilities and are innately attracted to simple ways of understanding. We "simplify with a vengeance" and are, it seems, victims of our own limited mental biology. Examples of this type of direct cause-and-effect thinking are reflected in our responses to complex problems such as housing shortage (encourage the government to build houses), food shortage in Bosnia or Somalia (fly in food), or low productivity on the production line (get the workers to work harder). This type of thinking is a problem for manufacturing modernization or for large-scale change efforts, because changes lag considerably behind actions.

8. *The principle of leverage,* or "small changes can produce big results." Senge argues that in tackling a difficult systems problem, we should be trying to locate leverage points so that minimal effort will result in significant improvement. Unfortunately, high-leverage solutions tend to be nonobvious ones as well, with the problems and solutions not closely linked in time and space. An example of using leverage might be changing the rewards system. As will be discussed later in this chapter, Xerox was able to exert a powerful influence on management behavior by promoting only those who had clearly demonstrated competence in total quality management. Employee empowerment and participation are also forms of leverage because managers actually increase their own power and reduce resistance to change by using these approaches.

9. *There is no blame.* In this error of system thinking, we tend to blame outside causes for problems. The outsiders might be competitors, the government, other workers, managers, or our families. Our belief is that these people have done something to us. However, in systems thinking the cause of virtually all problems is not external, but built into the relationship between "you and them." You are part of problem. Blame is an important issue in that assigning blame to someone else involves denial of our own responsibility and is likely to annoy those we blame. Whenever a problem is blamed on an external person or object, one may immediately assume that the cause of the problem is not understood and that the complainer does not see their part in the problem. In organizations, we often see one group blaming another—managers blaming subordinates, and subordinates blaming managers. At home, spouses blame each other or the children, or vice versa. The bottom line is that detecting blaming behavior should be a red flag for us to take a broader look at the issues.

Senge et al. (1994) described the tendency to blame as a "reactive," or victim, orientation that is often encouraged by the failure of organizations to allow active participation by members. Employees in such organizations feel "beset by a different level of gods and fates—the demands of customers, competitors, and employees" (p. 228). Those assuming the victim role, whether in work or nonwork situations, tend to keep their defenses up, avoid responsibility, and fail to maintain accountability for their actions. At the opposite end of the spectrum, other organizations have a "creative" orientation in which members are encouraged to achieve their own visions for the future and in which outstanding performance is rewarded. However, Senge found this orientation also unfulfilling, as persons with the creative orientation tend to become highly individualistic and concerned only for themselves. The culture of such organizations was characterized by a fast-pace, vicious competition among members, long work hours, and—ultimately—exhaustion. Senge's preferred alternative was described as the "interdependent orientation," a compromise in which individual goals are set and achieved, but within the context of a win–win orientation to other members and a strong identification with and commitment to the overall organization.

Senge's points about systems thinking and the problems people have with understanding and changing systems are highly relevant to the present discussion of managing system change. Unless we are able to develop strategies that avoid the kinds of problems he discussed, our efforts are likely to be in vain.

Faulty Strategic Decisions

Lawler (1991) developed an analysis of some of the key strategic decisions that must be made by organizations undertaking change efforts based on a system theory orientation. He pointed out that many seemingly sound strategic decisions prove to be flawed when considered in terms of the entire system. He stressed that strategic decisions regarding organizational change must be made only after considering the impacts throughout the entire system. Otherwise, systems energy is generated that serves to maintain the status quo across all organizational elements and to resist and reject efforts to change the system. The problematic strategic decisions examined in his analysis included the following.

1. *The decision to use experiments.* The decision to try out a new way of functioning in a small part of the organization appears, at first glance, to offer the opportunity to learn from experience and from mistakes and to apply these lessons to the entire organization. From systems theory and the notion of internal system congruence, however,

certain problems can be predicted. First, the small-scale experiment may give a misleading impression of how the change would work out in a large-scale application. What might work well in a small-scale test might not work organization-wide, and vice versa. Second, just as intrusive foreign material is attacked and rejected by a living organism, so will organizations tend to reject the "foreign material" of an experiment. The practices involved in the experiment will inevitably conflict with management systems and with existing policies and procedures. Failure of the experiment is a highly likely result. Third, because participants know that the change is only an experiment, little commitment to the new approach is built up. Fourth, even if the experiment proves to be a success, other organization members will have no sense of participation in the change and will tend to resist adopting the change. We would like to add a fifth possible problem with experiments: Apparent success may simply be due to an occurrence of the Hawthorne effect. The special attention given to the experimental group—independent of any true improvement or value—may motivate those involved enough to favorably bias the outcomes.

Overall, small-scale experiments were recommended only for cases where there was no real alternative and where structures could be established to protect the experiment from the rest of the system. Large-scale experiments involving changing a single aspect of the organization (for example, management style) also were not recommended, because systemic pressures for congruence would tend to reject the new approach. Pilot programs involving members across the organization and which, at the outset, were clearly intended for wide-scale adoption were suggested as a better alternative to experiments.

2. *The decision to maintain congruence of change processes and end state.* Compelling arguments may be made that the change process must be consistent with the desired outcomes. For example, it can be argued that adopting a participative approach should be done in a participative fashion. However, because a number of aspects of the system would need to be changed (for example, performance appraisal, selection, training, and so on) and because midlevel managers (who have excelled in the existing system) may be unwilling or even unable to support participative approaches, is it unrealistic to expect a participative

implementation system to work. Rather, clear vision and direction from top management regarding the implementation will be necessary. In fact, the change process may need to be congruent with the *existing* system, rather than the proposed system, in order to be effective.

3. *The decision to use a lead–lag strategy.* Because a large-scale change effort may involve modifying all of the major organizational systems, it would make sense to change these systems one at a time. One change effort could be planned and installed without unduly overloading the system. A second change effort would then be initiated, and so forth, until the entire system change had been accomplished. Unfortunately, the problem with the lead–lag approach is again system congruence. Each change would tend to unbalance the system, resulting in widespread resistance and pressure to reject the change. The simultaneous change of all parts of the system was seen as a more viable approach. Lawler offered a number of suggestions to support a simultaneous change strategy, but felt that "considerable work would be needed to develop such a model" (p. 263).

4. *The decision to blame and abandon the old system.* Much of the current change literature stresses the need to motivate people to change. A common approach has been to identify problems with the current system and develop comparisons with successful organizations (that is, benchmarking). The problem with this strategy is that it may arouse defensiveness in organizational members who have been involved in developing or using the old ways of doing business. Members may feel personally criticized and tend to respond with justifications rather than with change. Lawler suggested that the focus of change efforts should be on the advantages of the new system and not on problems with the existing one. Emphasizing the current system's good points while identifying changes in the business environment that require change would serve to defuse defensive behavior and help members feel good about the past.

5. *The decision to rely on programmatic change.* Arguments in favor of programmatic change—that is, using off-the-shelf change programs—include capitalizing on learning in other organizations, implementing established systems, and using existing manuals and training programs. However, such programs tend to have only minimal impact

on employee acceptance and commitment. The resulting change may be of cosmetic value only and may be quickly abandoned. Consequently, the use of programmatic approaches was not recommended. Rather, Lawler recommended giving each organizational unit the option of using off-the-shelf programs as part of its change effort, thus, enhancing participation and commitment. Regardless of the approach used, these units would be held responsible for implementation of the change.

6. *The decision to use either bottom-up or top-down change.* Many writers have stressed the value of bottom-up change, as this strategy results in high levels of participation and commitment. Others have argued that without top management support, change efforts turn into guerrilla warfare. Grassroots change leaders subsequently experience more or less severe forms of retribution from management and the change effort fails. Change driven by the top, however, disregards employee buy-in and commitment and generates considerable resistance to change. Lawler argued that *both* top-down and bottom-up change are needed for success. Top management should plant the seeds of change and can do everything possible to support and encourage the changes taking root (the so-called "agricultural model" of change) but will not be able to control what happens at lower levels.

7. *The decision to conceal the difficulty of change.* Another strategic decision involves the extent to which an external or internal consultant should fully inform organizational members of the degree of difficulty of the change and the amount of time and energy that will be required for implementing the change. The client organization may, if fully informed, be "scared off" by the reality confronting it. Based on strong ethical considerations, Lawler concluded that the organization must be given a full disclosure up front about the anticipated difficulties.

Critical Success Factors Derived from Failures

The previous review presents a number of different ways of looking at the success or failure of organizational change efforts. These writers displayed remarkably little disagreement in spite of the very different perspectives that were used. Instead, each writer seemed to focus on certain aspects of change that were problematic. In an attempt to integrate their concerns, the following list of "critical success factors" is offered, derived from themes summarized in the previous section.

Large-scale change must, in order to maximize the likelihood of success, incorporate the following critical success factors.

1. *Rely on a systems approach.* The organization must be seen as a complex system that, because of its complex and multifaceted nature, has a tremendous amount of inertia and can generate considerable energy to force a return to equilibrium when change efforts are initiated. Changing the system means not only changing diverse units across the organization, but also changing the standards, reward systems, planning processes, budgeting processes, information systems, and so on.

2. *Establish a strong results orientation.* The change process must incorporate clear expectations of real improvement in work processes and output and provide ways of measuring these improvements. Empirical results provide feedback and motivation to those responsible for the improvements.

3. *Require accountability by line management.* Accountability aimed at top management or support organizations will not work.

4. *Rely on top management for clear vision and direction.* While top management can support change, it cannot directly control change efforts. The job of top management is to provide the needed vision.

5. *Incorporate organizational learning.* Experiments (including use of prototypes and simulations) should be used in the process of achieving improvements and the results should be documented and broadly communicated. Benchmarking of successful efforts in other organizations should form a part of organizational learning.

6. *Incorporate results-oriented training.* Training should have a specific performance objective and should be provided only on an as-needed basis to implement changes. Generic, off-the-shelf training probably will not be of real value.

7. *Create simultaneous, organization-wide, and across-the-board changes.* Systems tend to stifle and reject isolated efforts—the changes must be broad in scope.

8. *Build the change process around the current system of "how things are done."* For example, in a bureaucratic organization, initiating empowerment or participation may need to be couched in terms of policies, procedures, and performance appraisal changes. Use all possible levers from the existing culture.

9. *Focus on positive aspects of the current system and on external factors.* Do not arouse unnecessary and harmful defensiveness by criticizing the current system.

10. *Rely on both a top-down and bottom-up approach to change.* Top management provides vision and direction, middle management is held accountable for ensuring that change takes place, and workers are empowered to initiate and implement changes to work processes from the shop floor level.

11. *Maintain flexibility.* Focus on achieving the vision while avoiding excessively rigid and detailed plans and strategies for getting there.

The critical success factors of managing change may be seen as benchmarks with which to evaluate change efforts. While no claim is being made that these factors represent all of the aspects of successful change, the success factors should at least provide considerable insight into whether a given change program is proceeding in an optimal direction.

The success factors can also serve the role of providing a framework with which to develop and/or evaluate a comprehensive change-management process. The goal of the next section is to use the critical success factors to review several of the leading change-management approaches, to evaluate these approaches in terms of likely effectiveness, and to provide a recommendation regarding a sound change-management model.

Use of Critical Success Factors to Evaluate Change-Management Models

Given all of the problems that have been identified with managing change, the natural question at this point is, "What is the bottom line on change management—what approach seems to work the best?" In order to help answer that question, a number of change models were evaluated in terms of their consideration of each of the 11 critical success factors identified in the previous section. The change programs included in this analysis were the following:

- Programmatic change approach (primarily involving off-the-shelf training)
- Kilman's "completely integrated" approach (Kilman 1989, 200–228)
- The congruence model of Nadler and Tushman (1977, 155–76) and Nadler (1989, 495–508)
- The technical/political/cultural (TPC) strand model of Tichy (1983)
- The sociotechnical systems design model of Pasmore (1988)
- The task-aligned organization approach of Beer, Eisenstat, and Spector (1990)
- The organization innovation model of Van DeVen (1993, 269–94)
- The high involvement model of Lawler (1993, 172–93)

The general conclusion of the review was that, with the exception of the Beer, Eisenstat, and Spector (1990) task-aligned model, the change programs addressed few of the critical success factors. The generic programmatic change approach, in fact, addressed none of them. The remaining seven programs all incorporated a strong systems-theory orientation and provided a comprehensive approach to addressing diverse parts of the organizational system. All programs incorporated the need for a top management vision. In addition, most of the models relied on a top-down change strategy—failing to link top-down with bottom-up change drivers. Top management, external consultants, and some type of transition team were expected to drive the change process throughout the organization. The programs in general failed to focus specifically on improving work and work effectiveness. Pasmore's (1988) sociotechnical model, although containing a strong work orientation, advocated work-design principles that do not match current industrial engineering approaches to work process improvement. Other models, such as Kilman's (1989, 49–72) devote considerable attention to culture change. Not only do the approaches to culture change seem simplistic, but I seriously question the value of attending to culture in the first place. I believe, rather, that achieving task alignment and getting the work processes and the production system in order will lead directly to the desired culture.

In my view, the Beer, Eisenstat, and Spector (1990) model was by far the most work- and results-oriented approach to change management available. The task-aligned model addressed all of the critical success factors, with the possible exception of focusing on positive aspects of the current system (number 9). Improving work processes and productivity are the key goals, rather than focusing on social or cultural issues. Beer's (Beer, Eisenstat, and Spector 1990) model included both top-down and bottom-up change drivers. He relies on top management for assessment, vision, and direction. He advocates a transition team for communications and coordination and on empowered employees to drive changes in their own work area from the bottom up. Further, the Beer approach builds in accountable midlevel managers to support change efforts in their units, offering a powerful strategy for dealing with midlevel management resistance. Overall, the model provides a powerful and compelling approach to managing the changes that are needed to initiate agile manufacturing. The next section outlines components of the model in some detail.

Creating the Task-Aligned Organization

Beer, Eisenstat, and Spector (1990) stress that using the task-aligned approach would create the coordination and teamwork across the organization necessary to revitalize the system, develop high levels of commitment needed to sustain the change efforts, and support the development of the new competencies needed to operate the new system. The ultimate goal of task alignment, as discussed in some detail in the previous chapter, is coordinating activities of all employees toward achieving the organization's competitive tasks (that is, the organization's strategic goals).

A key theme of task alignment is that the pursuit of self-interest by any single department or group results in suboptimization of performance for the overall organizational. For example, if the human resource function pursues its own goals, however well-intentioned, it will hurt production. This could occur by requiring better documentation of selection decisions, developing detailed policies and procedures for hiring and promotion decisions, developing a world-class affirmative action system, revising and refining the performance appraisal system, and so forth. While any of these actions could well improve the functioning of the human resources function, in all likelihood they will add no value to bottom-line production of the organization. Almost certainly, these activities will either actively interfere with production (perhaps by lengthening the time required to hire a new worker) or indirectly interfere by using scarce resources for nonproduction activities. The only way to avoid the suboptimization effect is for human resources to consider the production system as its primary customer and to take only those actions that serve this customer directly.

Similarly, other functions—product development, maintenance, quality control, production planning, and other departments—are also expected to adopt a support role in relation to the production system. The activities of these functions may have to change dramatically under the task-aligned vision. Thus, a broad and effective change management approach is critical to success.

The basic steps of the Beer, Eisenstat, and Spector (1990) change management model are provided in Figure 2.1. The model has been modified, primarily to embellish details within each step. Step four of their model, which reads, "Spread revitalization to all departments," has been broken up into two pieces—the production system and the support organizations—for clarity. I have renamed the last step "Continuous improvement," compared with their last step of "Monitor and adjust strategies. . . ." The approach outlined here involves the following main components: problem diagnosis, development of a vision, recruitment of broad management support for the vision, the initiation of work team–level performance improvements, and the creation of an environment where continual improvement is expected. The specific steps are described in greater detail in the following paragraphs.

In addition, a case study is provided in the next section to indicate more specifically how a successful organizational change process

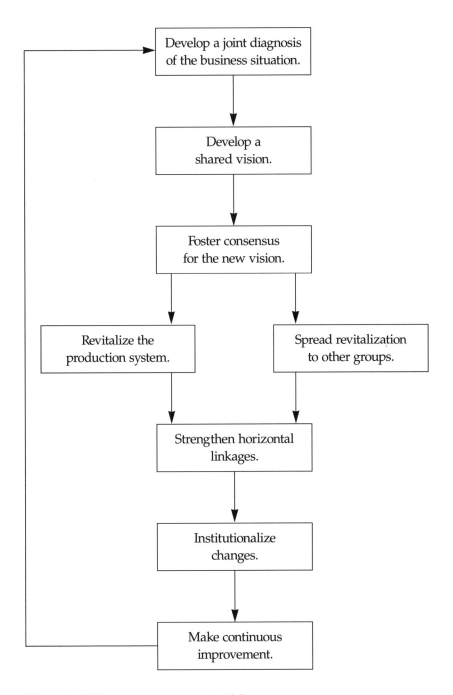

Figure 2.1. Change management model.

might work. The case presented is that of Xerox Corporation's transition to a high-quality and customer-oriented company in the late 1980s, based on the book by Kearns and Nadler (1992), *Prophets in the Dark*.

Step One: Develop a Joint Diagnosis of the Business Situation

In step one, top management conducts a strategic assessment of the organization in relation to its business situation or environment. The environment may include customers, competitors, regulatory agencies, suppliers, economic conditions, unions, transportation systems—a wide variety of organizations and conditions with whom the organization must "live" in the course of doing business. Organizations must develop a fit with their environment if they are to be effective and prosper (Pasmore 1988). The analysis must identify potential or real problems with members of the environment, such as problems with quality or customer service, declining market share, lagging productivity, inferior technology, and so forth.

The strengths and weaknesses of the organization itself need to be included in this analysis process. For example, the core competencies (Prahalad and Hamel 1990)—that is, the key business areas—should be identified. The analyses should answer questions such as, Who are the customers? Where does the organization stand in its ability to compete successfully? What are the key environmental forces and factors that must be reckoned with (that is, regulations, technological developments, economic conditions)? The organization must determine where it stands in relationship to its competitors, customers, and markets. This diagnosis results in a broad understanding of the competitive situation.

In addition, a more microlevel analysis is done to diagnose the functioning of the production system. To do this, a steering committee is appointed and charged with conducting a more detailed assessment of production functioning. The goal of this analysis is to determine the extent to which the organization is task aligned, as described in the previous chapter, and to develop a high-level diagnosis of the production system. Thus, while the production system will be the primary focus, the functioning of critical support groups is also to be considered. The steering committee is likely to require initial training in agile manufacturing concepts in order to properly conduct these analyses. The steering committee may also wish to conduct benchmarking analyses of target organizations, such as highly successful competitors, to identify world-class standards and learn from the experiences of these organizations.

The results of the business-oriented analyses of upper management and the production-oriented analyses of the steering committee are then communicated broadly throughout the organization through large meetings, open forums, and publications. The inclination of management may be to keep the findings secret, particularly if a number of troubling conclusions are reached. However, the acknowledgment of problems and the need to change is critical to prepare the members of the organization for the future. The greater the degree of change needed, the more intense the resistance to change is likely to be (Senge 1990). Further, the more successful the organization has been in the past, the less willing it will be to give up old behaviors (Pasmore 1988).

Step Two: Develop a Shared Vision of How to Organize for Competitiveness

As a result of the business situation and task-alignment analyses, upper management develops a statement of the mission of the organization, defining the business it wants to pursue and a vision of the world-class organization that is needed for success. The vision includes appropriate notions of agile manufacturing, world-class quality, customer responsiveness, and complete task alignment. (For detailed guidance regarding the visioning process, I suggest referring to *Visionary Leadership* by Burt Nanus [1992] or *The Leadership Challenge* by James Kouzes and Barry Posner [1987].) The mission and vision lead directly to formation of an organizational plan for improvement.

The steering committee also develops a vision statement—one that is at a more micro level—describing its best notion of the ideal task-aligned organization, in which every element serves its internal or external customers. The design includes an agile manufacturing approach and describes, in general terms, the ideal work and information flows. This model is then presented to upper management for review and acceptance. The goal is acceptance of the vision of a new way of doing business by upper management. As will be seen in the Xerox case study, obtaining such a consensus can be a difficult task that, if not accomplished at this point, will derail change efforts later.

Step Three: Foster Consensus for New Vision

In this step, upper management demonstrates its commitment to its own vision and the vision of the steering committee through announcements, publications, and meetings. Unlike the previous communications efforts, which focused on the business situation, this communications effort provides the vision for the future to the employees at all levels. The employees need to know the intended directions, as well as what will be expected of them individually. This is also a time to allow feedback from all levels and to demonstrate willingness to modify the vision, where needed, to correct for areas of oversight or error. Complete acceptance of the vision by employees should not be expected at this point. Rather, healthy questioning of and initial resistance to the changes should be anticipated. The role of management in these communications efforts should be to educate, inform, and listen—not to force or coerce. As noted by Senge (1990), change results in part from surfacing resistance and then dealing with the causes of resistance, not from attempting to impose changes by brute force.

A critical part of fostering consensus is what Beer described as the "velvet glove mandate" with middle management. Upper management communicates the vision to middle management, making clear that substantial autonomy will be given to midlevel managers for implementing the vision as long as they demonstrate commitment to that vision and do their best to support the transition. A clear contract or agreement is formally established with each midlevel manager. Where middle management agreement or commitment cannot be obtained, those managers are transferred or outplaced at an early date. This difficult step must be taken in a timely fashion. Midlevel managers have considerable ability to derail change efforts and may understandably feel that they have little to gain from the changes.

Without gaining their commitment up front, the change will not be successful.

Step Four: Revitalize the Production System

To this point, the change process has been largely top-down. Step four initiates the powerful bottom-up aspect of change. To start, a task force within the production system is established to coordinate and direct the revitalization efforts. Revitalizing is likely to involve substantial process reengineering and redesign. Everyone, including management, receives sufficient training to support achievement of the vision on a just-in-time basis. Task force efforts may include the benchmarking of successful competitors, although the ultimate goal is even higher standards, because the competitors will also be improving.

The task force initiates work process changes at the shop floor level that are guided by the vision of agile manufacturing and task alignment. Where existing work processes are very traditional and far from the desired state, a real revolution in production will need to be accomplished as production moves to a completely new paradigm of performance. For example, instituting a layout of the entire system by product, rather than by process, and implementing a pull system may be appropriate. Changes of this scope will need firm vision and direction from the transition team with support from the steering committee, upper management, and possibly outside experts as well.

At the same time, individual work groups must be included in the change process. The best ideas for improvement may well come from these employees. They must receive enough training to understand and support the revolutionary changes that are taking place in their area. Several possible approaches may be used. In Battelle's work with clients, staff have generally used the following approach (which is described in detail in a later chapter): (1) work process analysis, (2) education and training, (3) work process design, (4) planning, and (5) implementation. Xerox uses an approach called LUTI: learn it, use it, teach it, and inspect it. Regardless of the exact approach, workers must be properly trained in the elements of agile manufacturing, learn to stress quality and customer service, participate in the redesign of their work processes, and be involved in the implementation of the changes.

The work teams should be given substantial autonomy to improve their work processes within the framework of the new production vision once needed training in new production techniques is completed. Task force members or outside consultants must also be available to provide extensive support to these efforts as needed. While improvement activities may differ by work team, all changes must be driven by the vision of agile manufacturing and task alignment. Improvements may include improving work layout, reducing setup time, initiating statistical process control, gathering data on internal customer needs, initiating statistical process control, identifying ways of improving quality, and so forth. Performance measures (also discussed in a later chapter) will need to be developed and data collection initiated to track performance. Considerable experimentation during the improvement process should be anticipated, with failures as well as successes likely to occur. Depending on the size of the workforce, this step may require a considerable investment in time and resources, requiring perhaps a year or more to successfully achieve major work process improvement.

While work process improvement is underway, supervisors (and possibly midlevel managers as well) will need to be given training both to understand the technical change and to become aware of the new, more support-oriented roles that supervisors will need to assume. A later chapter on management issues addresses the supervisory role changes in greater detail. However, it is clear that because the work teams must have considerable autonomy to function effectively in agile manufacturing, supervisors must relinquish much of their (probably prized) decision-making and controlling authority. What is needed instead is for supervisors to provide support for work group problem solving and to facilitate working with other parts of the organization. The new role will require considerably different skills than before, such as greater interpersonal, communications, and facilitation skills. As with the velvet glove mandate, it may be that some supervisors are unwilling or unable to make the needed changes and will need to be transferred to more appropriate roles.

During the implementation of step four, the production task force monitors and supports the performance of the work teams. The task force reports progress to the steering committee and upper management on a regular basis, provides recognition to work groups for their successful efforts, and compiles and communicates lessons learned.

Step Five: Spread Revitalization to Other Groups

As shown in Figure 2.1, step five occurs in parallel with step four. The basic content of both steps is much the same. For each major support organization, a task force will need to be formed whose goal it is to implement task alignment and a customer service orientation. Major changes should be expected in the operation of these support organizations. The roles discussed in the chapter on the task-aligned organization should serve as the basic vision that guides these changes. As discussed in the previous chapter, these changes generally involve recognizing the internal customers, adopting a customer service orientation, and working with the production system to transfer to them key responsibilities (for example, preventive maintenance, quality control). It is also to be expected that some degree of downsizing will be needed for most support groups, following these changes.

The support organizations are likely to perceive themselves as the losers in the change efforts of step five. To optimize overall system performance, the support organizations will need to give up some of their previous autonomy, authority, and power. Because this is a lot to require, the reasoning behind the changes should be made clear. However, the velvet glove mandate was designed with the managers of support organizations in mind. Those midlevel managers of support organizations who cannot commit to the task-aligned vision must be transferred or outplaced as soon as possible. As with production, task forces are established within each organization to support the vision within that organization.

Training in agile manufacturing is provided to employees at all levels of these support organizations, just as in the production system. The training should be very practical and experiential in nature, with the assumption being made that workers will take the tools back to their work areas and begin initiating improvements. For example, it is anticipated that for each support organization, analyses will be performed on work flow and work processes to identify areas for improvement, initial improvements undertaken, measures of effective performance developed and implemented, and tracking of performance data initiated.

As in the production system, the task forces and managers should present results of their efforts to the steering committee. Where the results are inadequate, such as in overlooking the internal customer service orientation, revision and improvement of the approach will be required.

Step Six: Strengthen Horizontal Linkages

In step six, the steering committee works with each of the task forces created in steps four and five to focus on the chain of customers needed to create a fully task-aligned and agile organization. Additional communications efforts regarding the chain-of-customer vision may be needed to get the different internal organizational groups "on board." The forces for suboptimization, which include the desire for each group to maximize its own performance, should be analyzed and discussed. All groups must reach agreement to work for optimal system performance, rather than individual organization performance. The steering committee may develop customer-oriented performance measures to facilitate tracking customer-oriented performance within the organization. Where problems arise in certain organizations, the management of these groups may be asked by the steering committee to provide the training or specific direction needed to modify inappropriate actions.

It should be noted that if midlevel managers are committed to the new vision early in the improvement process, step six may prove to be little more than a formality. Individual work groups should have identified their customers in steps four and five and taken needed steps to provide the appropriate services.

The results of efforts to improve horizontal linkages should be monitored by the steering committee. Recognition should be provided to successful groups and organizations.

Step Seven: Institutionalize Changes

In this step, the steering committee analyzes progress toward agile manufacturing and task alignment and provides recommendations to upper management regarding any policy or structural changes needed to support the change effort. Proposed changes are to be driven entirely by business necessity and the need to increase overall competitiveness. Changes might include the adoption of broader, more flexible job classifications, revised pay and reward structures, revised criteria for promotions, or recommendations for more product-oriented or flatter organizational structures. Because these changes are to be driven by business necessity, employees and managers are likely to perceive them as fair and reasonable, and support them. For example, workers are likely to readily support recommendations for pay-for-knowledge so pay can reflect the new skills needed for job rotation.

The role of upper management is to review the recommendations of the steering committee and to implement the appropriate policy and structural changes. Some changes, such as those affecting union agreements, will need to be pursued appropriately to ensure legal compliance.

Step seven is one that may easily be overlooked, particularly by those with a strong technical orientation who have no intrinsic interest in organizational policies and procedures. In addition, these rules and policies may be extremely well entrenched and hard to change. However, as was the case with the velvet glove mandate, the hard issues have to be addressed in order for the entire effort to succeed.

One recent writer, Scott-Morgan (1994), has argued that the resistance to change encountered in an organization is mostly due to the existing reward and motivation systems. For example, resistance to a team orientation may simply reflect the fact that pay increases and promotions are still based on individual performance. What appears to be resistance is therefore simply rational behavior, given the system drivers and motivators. By simply changing these reward systems to ones that support the desired behaviors, change may be accomplished with relative ease and in a short time frame. Scott-Morgan's ideas directly challenge long-accepted views that real behavior and cultural change require a long, drawn-out process. If correct, his ideas suggest that organizational culture is a mere reflection of the sum total of all the operating reward systems in an organization—and, thus, culture is much more easily changed than culture theorists have proposed. Step seven is clearly a critical step in the overall change effort.

Step Eight: Continuous Improvement

Beginning with step eight, the activities of upper management, the steering committee, task forces, work teams, and individuals continue with the goal of continually improving organizational performance. The work processes, quality of output, and service provided to internal and external customers are monitored and expected to show continued improvements. Where this is not the case, the appropriate task force and/or work teams must assess and address underlying roadblocks. Throughout the entire organization, increasing levels of productivity, coordination and teamwork, commitment to improvement, and increased technical competencies should be anticipated.

A Case Study of Successful Management of Change

During the mid- to late 1980s, Xerox underwent a profound change in its performance, triggered by competitive pressures at home and abroad. The change process, lead by CEO David Kearns, has been documented in detail elsewhere (for example, *Prophets in the Dark* by Kearns and Nadler, 1992). The following section describes some of the key actions taken by Xerox during the change process, broken out by the steps of the change model just presented. The intention of this section is to provide concrete examples of behaviors taken for each step, as well as to indicate some of the problems encountered in a real-life situation.

Step One: Joint Diagnosis

David Kearns and a few of the members of top Xerox management were becoming increasingly aware that Xerox was facing serious problems in the early 1980s. For example, Xerox had lost a great deal of market share, going from a dominant market position to only 12 percent in 1984. Xerox was also facing a loss of income and profit as well as a low return on investment. Its return on investment was less than 10 percent, when 15 percent was common in manufacturing at the time. Further, the Japanese had specifically targeted Xerox as their next area of attack. In the United States, IBM and Kodak were beginning to develop their own copiers and were challenging Xerox.

Another serious problem facing Xerox was that of poor product quality. In 1979, for example, one of the copiers sent to market was not even built with a square frame, so toner leaked onto customer carpets. The printing quality was poor, the copier was continually breaking down, and paper jams were a nuisance. One Japanese commercial showed the copier with a repairman stationed in a nearby closet to be readily available for breakdowns.

There were also a number of internal conflicts. The East Coast staff tended to be financially oriented, very conservative, and focused on cost accounting. They blamed high costs for company difficulties. A number of key managers had been recruited from Ford specifically to improve Xerox's financial orientation and had done so with a vengeance. The more sales-oriented managers, who had previously dominated the organization, found themselves to be much less of a driving force. Further, a number of very creative and innovative researchers on the West Coast left Xerox virtually en masse due to resentment over loss of freedom and interference with their work by the financial staff.

At that time, Xerox had no real business strategy, other than the expectation that the company would continue being successful. In fact, other than Kearns and a few other managers, there was no widespread view that any problems existed. There was little organization-wide interest in quality improvement. Quality was seen as a manufacturing problem, with the general attitude being "just tell them to do it better." While there was a real fear that copiers might be rendered obsolete by new forms of technology, there was a general unwillingness to consider new approaches. For example, Xerox management resisted consideration of a "recirculating document handler," an existing technology that would have speeded up copying immensely, but it required the original document to be moved into the copier. The reason for rejecting the approach was that they feared damaging original documents and annoying customers, even though research showed this was not a problem.

To begin to communicate the problems he perceived and to initiate corrective actions, Kearns held a four-day meeting of Xerox managers from around the world. He arranged for demonstration of new products, but he also stressed in several speeches that Xerox was being outmarketed, outengineered, and outwitted by its competitors. His conclusion was that Xerox would never again have the market to itself, stressing the urgent need for change. Unfortunately, the managers were not convinced.

Faced with disbelief about his conclusions, Kearns sent a team to Japan to conduct benchmarking with a Japanese competitor, Canon. The team was shocked to find that Xerox was not even in the same league with Canon. The Japanese relied on much less inventory and overhead, maintained far higher quality, achieved greater productivity, and were improving their performance rapidly. Whereas the Japanese had 0.6 staff members per factory worker, Xerox had 1.3. Each Japanese unit cost two-thirds that of a Xerox unit. Because the Japanese were improving productivity at 12 percent per year, Xerox would needed gains of 18 percent to 19 percent for five years just to catch up. The Xerox team, alarmed by what it observed, attempted to communicate its findings back to Xerox headquarters. Although Kearns and few other managers heard the team's message, for the most part the report was discounted.

Comments. Much of the diagnosis of the business situation was done in a piecemeal and ad hoc fashion. Kearns and a few others began to realize that major problems existed, but were confused themselves by the complex events. While the worldwide meeting was an excellent attempt to communicate, the lack of systematic study of the issues allowed other managers to dismiss the conclusions. The benchmarking effort was appropriate and gathered critical data. It would have been even more helpful to create a steering committee and to begin researching internal quality and productivity problems. A key difficulty faced by Kearns was the lack of a model for diagnosing problems and for guiding change efforts. He was left in the highly stressful position of trying to figure out what should be done to save his company from problems he did not understand.

Step Two: Develop a Shared Vision

Kearns initiated additional benchmarking to get a better sense of how other organizations were functioning. Teams were sent to General Motors, Lockheed, and other U.S. corporations. David Nadler, an outside consultant familiar with organizational change (perhaps the best in the country), was recruited to help. Nadler benchmarked several plants run on sociotechnical principles, which were essentially operated by self-managed teams. At Cummins Engines (Jamestown, New York), for example, workers controlled their own schedules, made most of the production decisions, and were working under a pay-for-knowledge system. When these findings were presented to management at Xerox, many felt that Nadler was advocating communism. There was little support for employee involvement and empowerment. Most management ignored the benchmarking and production results, arguing that the company's problems were strictly in areas of cost cutting and cost control.

Kearns listened carefully to the benchmarking results and took it upon himself to adopt a vision of a new Xerox that was based on TQM. His vision included ultrahigh product quality achieved through employee involvement and commitment to excellence and a strong orientation toward achieving customer satisfaction. Kearns decided to become an active supporter of the new vision and brought Nadler in on regular basis for support in achieving change.

In addition, a transition team—called the "Gang of Eleven"—was created to support change. Team members were charged with learning about quality and communicating their findings broadly throughout the company. Through interviews with upper management, the Gang of Eleven found that only two of six top managers (whom they called "kings") supported quality, but seven of eight of the next level of managers (referred to as "princes") supported quality.

Comments. The turnaround of Xerox began at this time. Kearns correctly diagnosed the real problems afflicting Xerox as being related to quality and customer satisfaction. He developed a vision of the new total quality organization and lent his support to change efforts. Kearns brought in Nadler and created a transition team.

Step Three: Foster Consensus for the New Vision

Up to this point, support for Kearns's vision was quite limited. The goal of this step—fostering consensus support for the vision—was critical. Unfortunately, the Gang of Eleven failed to take strong action, a victim of lack of commitment, political game playing, and lack of capability of some team members.

The Gang of Seven was then formed, which included Nadler. The new team developed a 35-minute tape and slide show on TQM, which it took to the upper management under the guise of asking for assistance in improving the presentation. These meetings resulted in a good sense of each manager's position regarding quality. The team sent books and articles on quality to these managers, often underlining or marking key passages so that the entire book or article did not have to be read. In addition, the managers were sent on benchmarking trips to see the reality of the competitive situation. Well-known speakers on quality were brought in for presentations, including W. Edwards Deming, Philip Crosby, and Joseph Juran. However, it became clear that none of the quality gurus knew how to orchestrate the needed organizational change. Kearns published what was called the "Blue Book" describing Xerox's mission, the vision of the new organization, and the change that was needed. As a result of these efforts at consensus building, many of the members of upper management were brought on-board regarding quality.

The next idea was to require top managers to answer individual customer complaints one day a week. Managers had to answer these calls even if they were in important meetings. Managers quickly became aware of the depth of quality problems present. During this period, Xerox stock continued to plummet. The former arrogant attitude of management gradually was lost, and the severity of Xerox's situation became apparent to all.

Comments. The transition team and Xerox used a number of creative approaches to try to build consensus for the new vision. Considerable patience was demonstrated in these efforts. Unfortunately, lacking a strong competitive analysis (which should have been completed in step one), the team had a difficult time convincing Xerox managers of the problems. Further, considerable delay was suffered because of failure to implement the velvet glove mandate recommended for this step. Kearns expected people to change of their own accord and was unwilling to be confrontational. Contrary to his expectations, managers were not convinced by presentations, benchmarking results, or other data. Kearns later expressed regret that those who did not support the change effort had not been replaced early on.

Steps Four and Five: Revitalize the Production System/Spread Revitalization to Other Groups

Kearns and his transition team then initiated training in total quality for all of the 100,000 Xerox employees. They were aware of the tendency of training to *not* transfer back to the workplace and adopted a training approach called the LUTI model: learn, use, teach, inspect. The training started with top management and cascaded down into the organization. It was based on "family group training," in which entire work teams went through training together. Each attendee also had to serve as a facilitator in the training of his or her subordinates. This design assured maximal organizational support for implementing changes based on the training.

The training design included 48 hours of instruction on total quality management. It included a vignette describing real-life Xerox quality problems, showed a Xerox team meeting to solve a copier problem, and required work teams to identify their internal and external customers. A six-step problem-solving approach was presented, which required gathering statistical process data, avoiding quick fixes, and evaluating alternatives. The problem-solving approach was then used in simulations requiring the trainees to diagnose and correct production-related problems. The training also incorporated process skills modules regarding treatment of coworkers.

Overall, the training was highly experiential. Top management demonstrated strong supported for the training effort by actively participating and by providing the needed resources. When one manager halted the training in his department to avoid paying the training costs, he was promptly removed from his position.

The training not only provided direct performance improvements to the company, but created other benefits as well. For example, during the training, several staff groups found that they could not identify a customer for their services or they claimed to be their own customers. These groups became targets for elimination. In addition, the training revealed that headquarters functions were often in the role of providing information, resources, or materials to local plants. Thus, many headquarters functions were actually suppliers, rather than customers. These headquarters functions were forced to make major changes in their perceptions of their roles in the organization and assumed a more customer-service orientation.

As a part of evaluating training effectiveness, the transition team initiated employee surveys to see if managers were actually using the training. The surveys began to document work process improvement efforts, improve quality, and enhance customer-service orientation.

Comments. The training necessary for revitalization of Xerox performance was thoughtfully developed and provided. The critical top management support was available. The usual problems that result from a programmatic training approach were deftly avoided due to the highly practical and experiential nature of the training and because training was conducted for intact work teams. The velvet glove mandate (recommended for step three) was adopted at this time, with Kearns admitting it should have been done far earlier. The revitalization was directed at not only the production system, but at the support functions as well. All were expected to show improvements.

Step Six: Strengthen Horizontal Linkages

Step six was accomplished as the internal chain of customers began to be identified. The company-wide quality training, described in the previous step, proved to be a major driver. Both internal and external customers were identified and efforts were initiated to improve the customer-service orientation. By focusing on the chain of customers, the previously impenetrable functional stovepipes began to dissolve, to be replaced with a smooth flow of products and services across organizational boundaries. Functional groups began to see themselves as partners in production, rather than as competitors for resources.

Step Seven: Institutionalize Changes

A number of policies were changed at this point, including those for promotions. Under the new rules, in order to be promoted to a supervisory position, candidates were required to demonstrate their competence in quality improvement. For promotions in the management ranks, candidates had to demonstrate that they were *role models* for quality improvement. These new requirements for promotion proved to be powerful motivators of behavior changes.

Another institutional change was the requirement that all departments benchmark their performance with world-class organizations— not only with Xerox's competitors. Employees received training in benchmarking to support this effort. Eventually, every process was benchmarked. As a result, benchmarking became almost a trademark of Xerox corporate culture.

At this time, top management reviews of the quality performance of different Xerox organizations were formally initiated. These reviews lent further credibility and provided additional motivation in support of quality improvement. The message of the importance of quality was broadly conveyed. People realized that quality was not just another quick-fix that would eventually go away.

Comments. Promotion policies, benchmarking, and quality reviews were significant and effective steps to take to institutionalize the quality improvement process. Ongoing efforts, such as regular customer satisfaction surveys, were to continue. Additional actions could also be taken, such as instituting pay-for-knowledge, broadening job descriptions to support job rotation, or removing entire layers of management.

Step Eight: Continuous Improvement

Efforts were taken to ensure that quality and customer service continued to improve. For example, work teams posted charts in their work areas showing improvements. Xerox began to demand better quality from its vendors. A vendor certification program was set up, in which vendors had to display high quality and continuous improvement to be included. Employee teams began to visit customers to check their level of product satisfaction and to discuss problems. Customers were included as a source of input into product improvement efforts. The massive customer satisfaction survey process was linked even more closely with quality improvement efforts. The levels of customer satisfaction with each product and service were carefully tracked, and potential problems were identified and addressed. The times needed to respond to customer complaints and requests were measured, strategies were developed to provide quicker responses, and improvements in response time were charted. Cross-functional design teams were formed to include customers, design engineers, manufacturing engineers, and others in the design process. The new design teams resulted in improved product design, reduced design time, and improved manufacturability.

Comments. Clearly the employees had learned their lessons well. The actions taken showed considerable initiative at lower levels in the organization, close attention to quality, and a strong customer-service orientation. These actions represented a major change in behavior from the way business had been conducted previously.

As a result of the change process, Xerox performance began a remarkable turnaround. The financial measures started to improve, with revenues increasing from $8.7 billion in 1984 to $13.6 in 1990. Income rose from $348 million to $599 million. The return on investment went up from 9 percent in 1987 to 14.6 percent in 1990. Xerox gained seven points of market share during this time (with each point worth about $200 million in income). In 1989, Xerox earned the Malcolm Baldrige National Quality Award. The turnaround at this point was seen by Kearns as "20 percent there," but was well on its way. Overall, the change process was seen as ongoing, but also as a major success—one that kept Xerox from going out of business.

Conclusion

This chapter has examined some of the factors behind the success or failure of large-scale change efforts, has reviewed some of the main models of organizational change in light of these factors, and has presented a model of change management that I feel maximizes the likelihood of success of the change program. The model was developed with the vision of agile manufacturing and a task-aligned organization in mind. It is a bottom-line and results-oriented model. The eight-step approach described in this chapter is built on the need for an assessment of the business situation, development of a vision for the future, gaining consensus for the vision, and empowering members at all levels of the organization to improve their work processes. The model is primarily a participative one, yet it builds on the need for directive actions in certain key areas (for example, dealing with resistance and unwillingness to commit to the effort).

It is my belief that the management of change in a large organization is so complex that without a road map such as the one provided in this chapter, the task will seem overwhelming. David Kearns at Xerox struggled through long periods of bewilderment, confusion, and frustration in large part because he was not working from a model of change. How much simpler his task might have been if a model of change had been available to him!

References

Beer, M. 1987. Revitalizing organizations: Change process and emergent model. *Academy of Management Executive* (February): 51–55.

Beer, M., R. A. Eisenstat, and B. Spector. 1990. Why change programs don't produce change. *Harvard Business Review* (November–December): 158–66.

Chew, W. B. D. Leonard-Barton, and R. E. Bobn. 1991. Beating Murphy's law. *Sloan Management Review* 32 (3): 5–16.

Kearns, D. T., and D. A. Nadler. 1992. *Prophets in the dark: How Xerox reinvented itself and beat back the Japanese.* New York: Harper Business.

Kilman, R. H. 1989. A completely integrated program for organizational change. In *Large-scale organizational change,* edited by A. M. Mohrman Jr., S. A. Mohrman, G. E. Ledford Jr., T. Cummings, and E. E. Lawler III. San Francisco: Jossey-Bass.

———. 1991. *Managing beyond the quick fix.* San Francisco: Jossey-Bass.

Kouzes, J. M., and B. Z. Posner. 1987. *The leadership challenge.* San Francisco: Jossey-Bass.

Lawler, E. E., III. 1991. Strategic choices for changing organizations. In *Large-scale organizational change,* edited by A. M. Mohrman Jr., S. A. Mohrman, G. E. Ledford Jr., T. Cummings, and E. E. Lawler III. San Francisco: Jossey-Bass. Material used with permission.

———. Creating the high-involvement organization. In *Organizing for the future,* edited by J. R. Galbraith and E. E. Lawler III. San Francisco: Jossey-Bass.

Majchrzak, A. 1988. *The human side of factory automation.* San Francisco: Jossey-Bass.

Nadler, D. A. 1989. Managing organizational change: an integrative perspective. In *Organization development: Theory practice and research,* edited by W. L. French, C. H. Bell Jr., and R. A. Zawacki. Homewood, Ill.: BPI/Irwin.

Nadler, D. A., M. S. Gerstein, and R. B. Shaw. 1992. *Organizational architecture.* San Francisco: Jossey-Bass.

Nadler, D. A., and M. L. Tushman. 1977. A diagnostic model for organizational behavior. In *Perspectives on behavior in organizations,* edited by J. R. Hackman, E. E. Lawler III, and L. W. Porter. New York: McGraw-Hill.

———. 1989. Leadership for organizational change. In *Large-scale organizational change,* edited by A. M. Mohrman Jr., S. A. Mohrman, G. E. Ledford Jr., T. Cummings, E. E. Lawler III. San Francisco: Jossey-Bass.

Nanus, B. 1992. *Visionary leadership,* San Francisco: Jossey-Bass.

Pasmore, W. A. 1988. *Designing effective organizations: The sociotechnical systems perspective,* New York: John Wiley & Sons.

Peters, T. J., and R. H. Waterman. 1982. *In search of excellence.* New York: Harper and Row.

Prahalad, C. K., and G. Hamel. 1990. The core competence of the corporation. *Harvard Business Review* (May–June): 79–91.

Schaffer, R. H., and H. A. Thomson. 1992. Successful change programs begin with results. *Harvard Business Review* (January–February): 80–89.

Scott-Morgan, P. 1994. *The unwritten rules of the game.* New York: McGraw-Hill.

Senge, P. M. 1990. *The fifth discipline.* New York: Doubleday.

Senge, P. M., C. Roberts, R. B. Ross, B. Smith, and A. Kleiner. 1994. *The fifth discipline fieldbook.* New York: Doubleday.

Tichy, N. M. 1983. *Managing strategic change.* New York: John Wiley & Sons.

Van De Ven, A. H. 1993. Managing the process of organizational innovation. In *Organizational change and redesign,* edited by G. P. Huber and W. H. Glick. New York: Oxford University Press.

CHAPTER 3

Creating New Systems
Lawrence O. Levine

Transforming the traditional forms and practices of manufacturing into an agile organization requires many changes, as discussed in chapter 1. The management of change approach presented in chapter 2 provides an overall framework for navigating this transition. However, to succeed in this transition often requires a more detailed and broadly shared understanding of the organization as a production system and how that system needs to change to become more effective and efficient. An operational vision helps provide this shared understanding by defining how the major processes that deliver value to the customer are executed, controlled, and planned—in short, how the processes hang together as a production system.

Developing, defining, and communicating this operational vision is often difficult, as illustrated by some of the limitations of the various tools and techniques described in this chapter. However, a step-by-step approach is described here (that is, triple diagonal modeling) for effectively and efficiently accomplishing this goal.

Why Is Understanding Your Business as a Production System Important?

An organization's production system can be defined as all of the processes that go into making the organization's product. A process is a series of activities that transforms an input into a desired output using people, equipment, and information to add value. Process improvement involves examining a process to see if its performance can be improved by eliminating steps, simplifying activities, and so on. Process reengineering, a concept made popular recently, involves totally redesigning a process by challenging basic assumptions about the way its activities are accomplished. Like any other process, a manufacturing process can be improved or reengineered. Likewise, an entire production system can be improved or redesigned as well. However, to begin improving or reengineering processes in a manufacturing organization without having an operational vision for the production system is dangerous. Why? Because a production system is more than a set of independent processes. It consists not only of processes on the shop floor, but processes that plan and control what happens on the shop floor. A healthy production system links these activities by feedback mechanisms to ensure that outputs meet customer needs effectively and efficiently. As shown in Figure 3.1, planning processes take customer needs and orders and translate them into production plans, as well as ensuring that resources are available where and when needed throughout the organization to meet objectives. Control processes monitor both the production plan and shop floor feedback, and give instructions and corrections to the shop floor processes. A clear operational vision that describes how all these processes work together is mandatory for an organization wanting to become both efficient and agile.

Production systems, like all dynamic systems of any complexity, often behave over time in ways that are difficult to comprehend. Symptoms like excessive inventory or missed schedules often have multiple causes. Moreover, symptoms may be far removed in time and location from their underlying causes. This separation can lead to an organizational culture based on invalid generalizations and superstitions about the causes of problems and how to get work accomplished (Senge 1990). Developing and communicating a common understanding of your production system helps overcome this cultural problem, and

- Improves alignment of performance measures within the organization

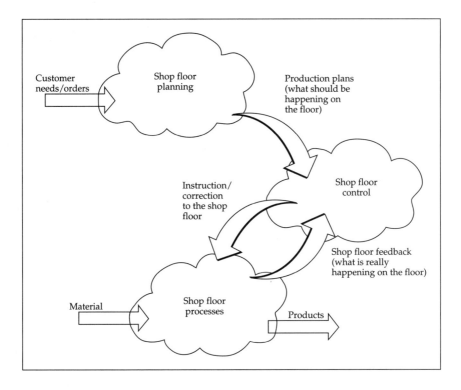

Figure 3.1. Operational vision of the production system.

- Minimizes functional infighting
- Improves consistency in business operational performance
- Provides greater sophistication in assessing and adopting (or rejecting) new improvement buzzwords and acronyms
- Provides better linkage between change management efforts and specific efforts to change processes and technology

The need for these improvements is described in the following sections.

Alignment of Performance Measures

Often, performance measures are inconsistent within an organization because of inadequate understanding of the production system. For example, for many years, production management was judged on maximizing machine and labor utilization while also responding to unexpected customer demands. Queuing theory indicates that these two objectives are contradictory (Blackstone 1989), and trying to pursue both objectives simultaneously frequently results in large backlogs and poor service. Because these criteria are inconsistent, the practical consequence is often a production system that oscillates between different symptoms as managers strive to improve one performance measure, which subsequently worsens the other (Senge 1990).

Functional Integration

In all production systems, the activities that support adding value (execution), ensuring the work is done efficiently and effectively (control), and identifying and providing the resources to accomplish the work (planning) all contribute to meeting customer needs. These activities usually cross functional boundaries and have important feedforward and feedback interactions between them. A frequent complaint heard in such circumstances is "Give me control of function *X* and I'd make this place work." However, believing in organizational fixes by themselves is almost always a mistake. The real issue is usually missing elements of control and planning or excessive cycle times in the processes or feedback mechanisms that prevent effective performance.

Consistent Operational Performance

Planning processes that are inadequate often generate production plans that are not achievable when created. Such plans should be based on the proven capacity of the organization and its resources, not issued as "stretch" objectives. The flexibility to modify such plans to meet shifting customer needs must be based on a realistic understanding of the cycle times to execute changes and a management discipline to limit the impact of these changes. Inadequate understanding of a production system can result in unrealistic performance expectations, constant micromanagement and redirection of effort, disappointed customers, and excessive costs (inventory and overtime). Ultimately, greater flexibility is achieved through better integration and cycle time reductions in execution, control, and planning activities.

Quicker and more consistent execution and feedback can have profound impacts on how the manufacturing organization works as a production system. Understanding and acting on this fact gives management real leverage to improve the business (Plossl 1991). The production system can rely less on forecasting and more on actual customer demand. Reduced cycle times and variability mean the production system is less likely to lag or overshoot demand in adjusting capacity and output. In general, cycle times and organizational slack (inventory, people, and so on) are directly related. As cycle times go down, slack is reduced. Moreover, the organization is better able to identify and correct quality problems, especially those that cross functional boundaries.

An Intelligent Consumer of Buzzwords

Throughout much of the 1980s, too many U.S. executives jumped from one buzzword to another (MRP, JIT, TQM, DFM, and so on) without a fundamental understanding of the impact of how quickly and evenly movement of material and information changes the nature of production systems. Just-in-time (JIT) was viewed primarily as a way to reduce inventory costs, but the link between reduced lot sizes and quicker feedback on quality problems was often overlooked. Cellular layout was viewed primarily as a way to reduce manufacturing lead times, but its impact on significantly reducing shop floor control complexity was often disregarded. Quality needed to be improved because the customer demanded it, not because reduced variability (consistent quality) was needed to eliminate the cushion of inventory and lead time that slowed responsiveness. Design for manufacturability (DFM) was a way to reduce or eliminate direct labor at assembly, without always seeing how reducing part counts and standardizing components could improve the speed of design and reduce the number of engineering change orders. Manufacturing resource planning (MRP) was seen as a way to manage manufacturing with information. However, shop floor complexity and delayed processing of inventory transactions often made the data in such systems inaccurate and outdated, leading to many failed implementations.

While all the various acronyms have merit, their implementation frequently is difficult because of a lack of integrated effort to speed up and smooth out the flow of material and information. When ad hoc, fire-fighting activities dominate an organization, it is a often a symptom of a production system with some role or activity that is not being addressed. For example, shop floor personnel often ignore production schedules that have not been based on realistic estimates of capacity. In turn, available capacity should reflect time spent doing regular preventive maintenance (PM), or PM should be explicitly scheduled along with production. When identifying these gaps, it is important to remember that people often have never been educated in the importance of fulfilling these roles (that is, how this fits into the bigger picture) or given the training to perform effectively.

Linking Change Management and Process Changes

It is important to realize that production systems—not functions or technology—are what consistently deliver value. Moreover, production systems are often the basis of sustained competitive advantage and customer loyalty (Stalk, Evans, and Shulman 1992). Management must ensure that the systems being created or modified are aligned with customer-defined needs. To do this, top management must share a commitment to a common operational vision that can be widely understood. This is not the same as a commitment to support organizational change. Management needs to understand the connection between these new systems and specific business needs, and to visualize how new roles, activities, or operational discipline support these needs and the functioning of the system. With this understanding, it becomes easier to prioritize process improvement and technology implementation efforts. Resistance to change is also reduced throughout the organization when people can see how a particular improvement fits into the big picture.

Why is it important to understand your business as a production system? It is important because becoming an agile organization is more than identifying a set of practices or technologies and creating a project plan for implementation. Getting from here to there requires creating and sharing a new conception of how the organization will operate. This operational vision must help all employees to conceptualize a new way of interacting across processes and functions. This vision must be linked to an understanding of how value is added for customer satisfaction. It needs to address in understandable terms how the firm will flexibly respond to changes in demand (both volume and product mix). Finally, the vision needs to serve as a framework for identifying and prioritizing specific improvement projects to ensure consistent movement toward strategic goals.

How Can You Understand and Communicate About Your Production System?

There are various techniques and methods used to understand your current production system and communicate how it needs to be transformed. Several approaches are highlighted here.

Computer Modeling

Computer modeling is one approach to creating and sharing a vision of the operation of the production system. Computer modeling has the ability to handle multiple feedback loops and can provide insights into system behavior that is often nonintuitive, identifying potential problems prior to implementing proposed changes and improvements (Forrester 1992; Gabriel, Bicheno, and Galletly 1991; Senge 1990; Senge and Sterman 1992).

Though current software packages are becoming easier to use, computer modeling usually requires the help of a staff expert or outside consultant to implement. Personal computer simulation software can be purchased for $1000 to $2000. More sophisticated capabilities and features can increase this cost significantly. Most software is based on approaches using either discrete event simulation or systems dynamics concepts.

Line management and hourly employees are likely to be suspicious of these techniques because of the difficulty in explaining the structure of the model and their limited familiarity with statistical concepts. As computer models become more sophisticated to more closely represent the "real world," they often become harder to modify and communicate to the larger audience within the organization. Both these attributes can limit modeling's role in supporting change management and identifying critical system performance issues requiring management decision making (Issacs and Senge 1992). In general, computer modeling is more useful as a tool to evaluate system design alternatives after agreement has been reached at a qualitative level on the new production system structure (Suri 1988, 118–33).

Human Simulation

Human simulation, or game simulation, uses people to play roles to illustrate how the structure of interaction and decision making in a system can influence overall performance. While gaming has been used for several decades as a management education tool, its use in educating a broader cross-section of employees seems to have increased to illustrate the difference between push and pull production systems (Denton 1990; Wu 1989). It is particularly useful in illustrating the impact of reduced work-in-process levels on production cycle time and identifying quality problems. This is especially important for firms still making the transition to lean production techniques (push to pull). Many employees and supervisors are suspicious of attempts to reduce safety stocks in front of their work areas (understandable, given that many are evaluated based on machine or labor utilization).

Human simulation is frequently effective with a wide cross-section of employees. It has the advantages of being easily learned, which encourages wide participation, and of being relatively inexpensive to design and conduct. However, it can usually deal with only a few important variables at a time and can bog down if the mechanics of conducting the simulation go wrong.

Human simulations usually take several hours to complete and require adequate space to both play the game and debrief the experience with the participants. Often the game is rerun several times to illustrate how changing key aspects of the game results in significantly different outcomes. Attempts to make human simulations more realistic and complex soon run the risk of boring or confusing the participants instead of enlightening them. The simulations seem to work best as a means to convey a qualitative understanding of a few key concepts that need to be understood before transforming a production system. Therefore, human simulation should be considered a tool for education and consensus building, rather than a system design tool.

Benchmarking

One way to create a common vision of a new system is for a group of individuals in an organization to benchmark the system of another organization identified as best-in-class (Camp 1989; Watson 1992). Benchmarking another organization can help overcome resistance that the new vision is impossible or unrealistic. It can open the minds of people to new possibilities, as well as provide specific ideas for details of the new system that may be hard to anticipate (that is, learn from other people's mistakes).

However, benchmarking can backfire as an approach to creating the new system vision. If adequate analysis of the current processes has not preceded the benchmarking effort, the visits to other organizations can degenerate into "industrial tourism." The participants may be ill-prepared to ask questions or absorb the lessons to be learned by the experience. Even worse, participants may become attracted to superficial aspects of the system at the benchmark organization and perceive them to be the key to success, without understanding how specific mechanisms fit together into a coherent, effective system. One way to avoid these pitfalls is to follow a structured approach to benchmarking.

Step 1: Select initial benchmarking area(s) and form benchmarking teams.

Step 2: Prioritize elements of benchmarking area and write a scope/problem statement.

Step 3: Describe existing system and identify problems.

Step 4: Establish performance measures to monitor improvement.

Step 5: Fix the problems with clear solutions.

Step 6: Identify potential benchmarking partners and best practices.

Step 7: Exchange information and visit selected sites.

Step 8: Develop short-term and long-term implementation plans and follow up.

In summary, benchmarking can be useful as part of a broader effort to create a new system vision. It needs to be preceded by an analysis of the current system to prepare those who plan to take the trips, as well as those left behind who have to rely on the findings of the direct participants.

Business Process Reengineering

Business process reengineering (BPR) is the latest acronym to promise significant improvement in organizational performance (Hammer and Campy 1993; Davenport 1993). BPR looks to achieve significant improvements through radical changes in the technology, people, and organization of key business processes that cut across organizational (functional) boundaries. BPR typically starts by identifying the key high-level processes or business streams in the enterprise. This list is then prioritized to focus on key areas to improve. Subsequent analysis generates a series of improvement projects that are aggressively managed. Given the scope and speed of the changes, BPR usually tries to support the activities with an underlying change management approach to deal with the various "soft" issues that might derail the process.

There are several characteristics that lend BPR to creating an operational vision as the basis for moving forward and clarifying how various improvement projects can fit together. BPR tries to take a systems view by identifying the key business processes of the organization. It also tries to prioritize improvement efforts based on their impact on external customers and key strategic objectives. It stresses the need to deal with people and organizational issues by explicitly addressing change management. Finally, successful BPR recognizes that even radical improvement has to be accomplished in manageable bites and usually limits these to one to three key projects per year.

Unfortunately for small to midsized manufacturers, BPR, as currently practiced, has several significant drawbacks. First, it is usually expensive and is more appropriate for large organizations that have both adequate internal resources and ability to hire consultants to work for extended periods of time. Second, the failure rate of BPR projects has been high. Smaller firms may not survive the cost of a failed BPR project. The analysis that happens at a high level in BPR may give little indication of how key processes fit together. More detailed maps often must be created to identify key problems, but this analysis can become a tar pit of effort and time. The current approaches to BPR give little guidance on how to create the "to-be" design.

If small to midsized manufacturers need to reengineer, they need to do it as quickly and inexpensively as possible. Clean-sheet approaches advocated by some reengineering proponents may be impossible for smaller companies to afford. Some reengineering approaches seem to suggest that setting impossible performance goals, doing a quick mapping of the as-is process, and learning to think "outside the box" is sufficient guidance and methodology for radical improvement. This gives little practical insight into how to make the reengineering team members or their superiors understand and communicate about the production system if they don't already have a common framework.

Summary

These various techniques for developing and communicating an operational vision have their strengths and weaknesses. For most small to midsized manufacturers, these techniques can be better utilized in conjunction with a practical low-cost approach to solving this problem that is described in the next section.

Triple Diagonal Modeling

Triple diagonal (TD) modeling is a technique to help quickly diagnose an organization's existing production system and to identify significant improvement opportunities in executing, controlling, and planning operations (Shunk, Sullivan, and Cahill 1986; Levine and Villareal 1993, 406–11). Because it addresses all three aspects of production system performance, TD modeling can provide a comprehensive framework to conceptualize how to improve both the manufacturing process and its support processes.

A TD model is a large drawing that maps the relationships between key activities in execution, control, and planning. It also includes information on resource consumption, cycle time, and quality. The drawing is constructed using the people within the organization. The completed drawing helps highlight where key performance problems exist, long cycle times are impacting cross-functional integration, and poor linkages or missing functions in the production system are getting in the way of meeting business objectives.

TD modeling has several advantages over other modeling techniques. First, it quickly does as-is analysis and then moves on to identify improvements. Analysis efforts can be tailored to the requirements of the organization by scoping the model to cover the few key activities or products that are critical to success. Second, creating one large diagram makes it easier to share the TD model throughout an organization, as opposed to the many linked $8^{1}/_{2} \times 11$-inch drawings used in traditional functional decomposition approaches to systems engineering. Third, it acts as a communication mechanism to share understanding about improvement opportunities that may cross existing functional/organizational boundaries. Finally, TD modeling acts as a vehicle to build a consensus on a prioritized list of improvement efforts that "hangs together" as an agenda for systemic changes in the production system and the improved integration of support functions. TD modeling is derived from ICAM definition language (IDEF 0)— also known as structured analysis and design technique (Marca and McGowen 1988).

Unlike most functional decomposition approaches, TD modeling starts by modeling the flow of material among major functional operations to produce the product or service (that is, starts at the bottom rather than the top). Figure 3.2, which shows a simplified TD diagram of a production system, shows several major functional operations along the bottom. Above this execution level, the model is expanded to incorporate information flows among functions that act to modify or control the execution functions. Above the control level, a third level of planning functions is added to describe how production is scheduled and how the execution and control functions receive the physical and information resources needed to meet that schedule.

A Remanufacturing Triple Diagonal Example

Figure 3.3 shows a TD diagram for a remanufacturing production system. This organization takes in (inducts) products that can be remanufactured, replaces worn out parts, refurbishes other parts, and then reassembles the product for return to service. On the execution level,

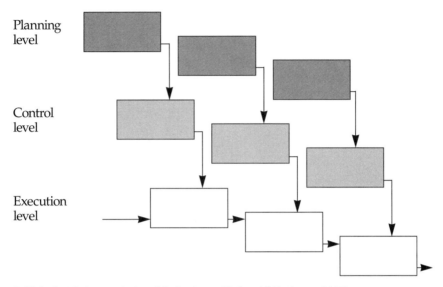

Figure 3.2. Triple diagonal schematic.

three primary activities are identified: disassembling the product (referred to as assets or cores), rebuilding components, and reassembling the final product. The control level shows three critical activities: managing the flow of products into the facility for processing, managing the inventory of parts (particularly monitoring the yield of refurbishable components), and scheduling the final assembly function. The planning level shows three key activities: planning longer-term production goals (particularly related to the forecasted availability of assets for remanufacturing), developing the engineering data required to remanufacture the products, and developing monthly production schedules for the entire operation.

The figure distinguishes the flow of material (solid lines) from the flow of information (dashed lines). Also noted within each activity are estimated worker hours expended each month. Where quality or yield problems are particular concerns (for example, the yield of repairable components from the disassembly activity), this is also noted on the diagram.

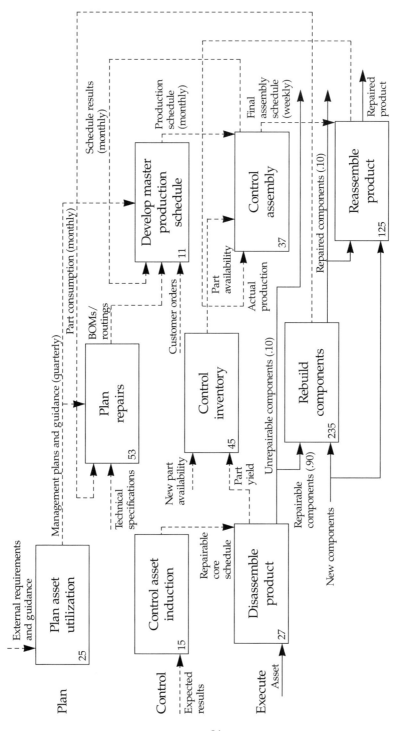

Figure 3.3. Triple diagonal of a remanufacturing production system.

This example is a simplified version of an actual application of the TD modeling technique. The original drawing contained about 75 separate activities, rather than the nine shown here.

The Eight-Step TD Modeling Process

Building a TD model is usually easier when a structured process is followed. Experience suggests following an eight-step modeling process.

1. Determine the basic flow of the product.

2. Determine the material flow.

3. Add control feedforward.

4. Add control feedback.

5. Add planning feedforward.

6. Add planning feedback.

7. Quantify the material flow.

8. Quantify the information flow.

These steps are described in more detail in the following sections.

The development of the TD model usually involves a core team of people. As the model is developed, additional people are brought in to validate the diagram and to add missing elements. In practice, it is often easier to create initial drafts of the diagram on large (4–5 feet wide by 10–20 feet long) pieces of paper using taped index cards and pencil. After the draft has gone through one or two revisions, inputing the paper diagram into a computer graphics package can make a cleaner and more easily duplicated drawing. One or two subsequent revisions are normally required to finalize the diagram. Keeping the TD model at a useful level of detail usually results in a diagram with 20–50 separate functions or activities.

Step 1: Determine the basic flow of the product. Before beginning the process, the purpose and scope of the modeling effort should be clear to ensure the project can be completed within time and budget constraints and the results are of value to the organization. This can often be achieved by selecting a key product family or critical production process as the focus. For example, a remanufacturer of helicopters chose to focus its modeling effort around the remanufacturing process associated with one key type of turbine engine. Improving the turnaround time for this engine was key to its competitiveness. In addition, the problems and potential solutions associated with this engine remanufacture were widely applicable across the organization.

Part of determining the scope is deciding on the boundaries of the system to be modeled (that is, the first and last activities that transform the product). If key processes are subcontracted to vendors, these should also be mapped. In general, the boundaries need to be broad enough to ensure that a focus on the end customer is not lost.

The basic flow of the product is mapped along the bottom of the diagram in what is called the execution level. The execution level includes those activities most directly associated with transforming inputs into outputs that are of value to the customer.

One of the key questions in mapping the execution level is achieving the right level of detail. Too much detail and the TD model becomes unintelligible. Too little detail and the TD model fails to provide adequate insight into the production system and its problems. One rule of thumb is to break the process flow where material movement or handling equipment is required to identify separate execution-level activities. For example, in the disassembly of a turbine engine, various subassemblies are routed to different specialized shops for further teardown and evaluation. This teardown and evaluation would be considered one activity in the execution level. If pieces in this subassembly are then routed to multiple areas of the facility for refurbishment (to be worked on shared process equipment), each of these would be considered separate activities also. In contrast, all the parts that were refurbished at the original disassembly location would probably be mapped as one common activity.

A good check to ensure that the activities are defined properly in the execution level is to examine whether the input of an activity appears to have been transformed at the output. For example, inspection of work by quality assurance specialists is not usually considered an execution-level activity (it is a control-level activity), unless the specialists become involved in any rework activities. However, the rework process itself should normally be mapped as a distinct execution-level activity.

Step 2: Determine the material flow. Inputs and outputs of each execution-level activity are identified. These can include purchased and fabricated parts, and scrap. Outputs of one execution activity that are used as inputs for another such activity are also marked on the drawing. During this phase of the mapping process, additional execution activities are sometimes identified and others may be combined.

Step 3: Add control feedforward. The modeling effort now moves up to the control level. Control activities monitor and control the activities in the execution-level. First, control activities are identified. Next, outputs from control activities that are used as controls to execution-level activities are noted.* Control inputs trigger or influence the transformation process of an execution activity (for example, a job order packet may identify the specific work instructions to be followed). Frequently, inspection by quality assurance specialists is mapped as a control-level activity as it can determine whether a part can move forward in the process or must be scrapped or reworked. Finally, outputs from one control function that serve as inputs to another control function are identified (for example, job order scheduling feeding the development of a work center dispatch list).

Step 4: Add control feedback. For control to be effective, actual performance must be compared to expected output, and then a determination of what (if any) corrective action should be taken must be made (as illustrated in Figure 3.4). Information outputs from the execution level (for example, production and scrap reporting) and the planning level (for example, production requirements) serve as inputs to control activities. In some cases, the output of a control activity may serve as an input to another control activity as part of the feedback process (that

---

*For those familiar with IDEF 0 or SADT syntax, this may seem confusing. TD modeling, like IDEF 0, uses control inputs that trigger or influence an activity. However, there is no distinction between types of activities in IDEF 0. TD modeling explicitly distinguishes between execution, control, and planning activities. Therefore, in TD modeling one can diagram a control activity whose output serves as a control input to an execution activity.

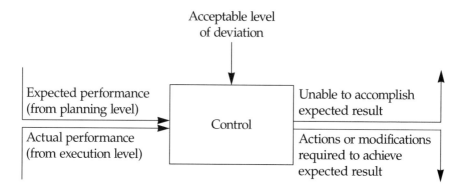

Figure 3.4. Generic control activity.

is, an evaluation of the requirements to rework an item can feed the rescheduling of a job order).

Step 5: Add planning feedforward. The modeling effort now moves up to the planning level. Planning activities ensure that adequate resources are identified and available to meet realistic production schedules and achieve other business objectives. First, planning activities are identified. Next, outputs from one planning activity that are used as inputs to another planning activity are added to the map. These data flows usually represent relatively stable data, like bills of materials and routing, that feed the generation of a master production schedule. Outputs of planning activities can also serve as control inputs to control activities. These outputs are typically schedules and revisions to previous schedules, and authorizations to adjust near-term capacity (for example, approval of overtime).

Step 6: Add planning feedback. The output of some planning activities can serve as a control feedback to influence another planning activity. A typical example (as shown in Fig. 3.5) is doing rough-cut capacity planning, which determines the infeasibility of a proposed master production schedule (MPS). This causes the MPS to be modified to fit within capacity constraints. Often, planning activities require feedback from lower levels to replan when operations cannot achieve original goals and schedules. While every effort should be made to get back on

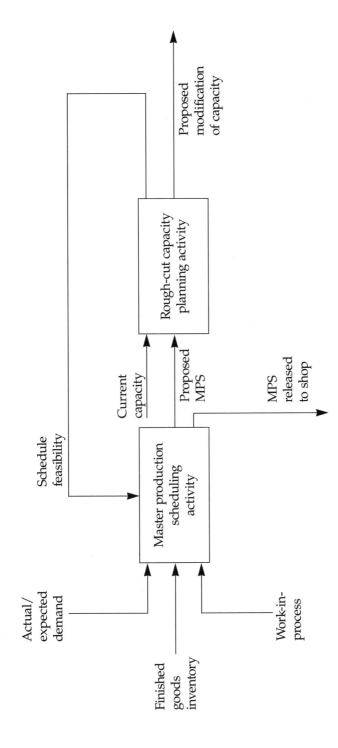

Figure 3.5. Rough-cut capacity planning provides feedback to generating the MPS.

89

plan, this is not always possible. This feedback may be from either control activities or execution activities.

Step 7: Quantify the material flow. Once the basic model is defined, it is selectively annotated with important quantitative measures of resource consumption and performance. This effort usually requires some analysis of records and/or questioning knowledgeable staff. The level of resource consumption (for example, worker hours or machine hours) and cycle time are determined for each execution activity. Any waiting time is usually included as an element of the activity cycle time data. Time devoted to moving the material can normally be ignored.

The quality of key inputs and/or outputs is also identified and annotated on the diagram. It is not necessary to identify all quality (defect) data; only those activities that generate significant scrap or rework. In practice, all these quantitative data are usually identified a bit at a time and are added to the map in several iterations as they become available.

Step 8: Quantify the information flow. Finally, the processing of information is quantified. This involves quantifying the resources consumed in each control and planning activity, determining cycle times for information processing and decision making, and evaluating the quality (accuracy) of the information flows. This information is also added to the model. Frequently, only estimates of cycle times for information are available. Time should be spent confirming the cycle times for those information processing/generating activities that have the biggest impact on the execution-level activities and/or on the customer. Similarly, estimates of resource consumption for the largest items should be confirmed first.

How Does a Company Use the TD Diagram to Build an Operational Vision?

Experience in using the TD modeling process suggests that several elements must be addressed to ensure it helps build a vision of a new operational system. Management must support the effort and understand the substance of the diagram, preferably as it evolves, at least after the modeling effort is completed. The modeling effort should be based on a cross-functional team that represents an organizational cross-section—both vertically and horizontally. The team must have members with adequate understanding of the basic flow of the operations. If the flow is particularly complex, the team should consider limiting the modeling to a few key products or components that are judged by management to be critical to competitive success. While the team has primary responsibility for creating the model, a consultant can often be useful when an organization first starts to pursue TD modeling. This consultant should provide several types of support. The consultant should be knowledgeable of the modeling technique and provide training on the modeling syntax to the group. The consultant should work with the team to ensure the model is kept at an appropriate level of detail. In addition, the consultant should provide education in the basic theory of manufacturing as a production system to create a common understanding and vocabulary among the modeling team. Once this basic framework has been presented, most people from the shop floor can understand the model and identify how limitations or gaps in control and planning activities impact operational performance.

Presenting the results of the modeling effort to the organization is critical to building the new vision. In practice, combining the presentation of the TD model with a human simulation based on a simplification of the organization's production environment has proven very effective as a communication tool and consensus builder. Combining this experiential learning with some overview education on the key technical areas needing improvement can help fill gaps in management and staff understanding and present a more realistic preview of the new skills and techniques the organization will have to adopt. This type of awareness training can be obtained from consultants, tapes, or reading and discussion.

The model should form the focus of a strategizing effort to identify and prioritize improvements that will impact organizational performance. These improvements will typically focus on speeding up the flow of material and information throughout the production system and filling gaps in the structure of execution, control, and planning activities. A good rule in prioritizing improvement opportunities is to look for those areas or processes with the largest cycle time that impact the customer or the cash flow of the organization (for example, order fulfillment).

Creating the New System and the TD Model

After the improvement projects have started, additional data collection and detailed design work will usually be required. These efforts may find traditional functional decomposition approaches to analysis and documentation useful. However, the organization will often find it useful to retain an up-to-date version of the TD diagram to serve as the integrating framework for the new agile system.

As improvement efforts move forward, the TD diagram will have to be modified. However, these changes will probably be modest. Some new activities may be added, functions combined, or connections between activities modified. Cycle times should be revised to reflect progress. The basic structure of the model will probably remain, unless significant changes are made to the underlying production technology.

The TD diagram and human simulation that were presented to management can be used throughout the improvement efforts to communicate throughout the organization. These techniques should be considered part of the tool kit for managing change on the road to agility.

Conclusion

Managing the transition to agile manufacturing has both hard and soft side issues. Creating an operational vision of how the current production system needs to be changed and communicating it throughout the organization is something that cuts across these issues. This common vision creates the framework for the difficult work that must be tackled to become agile. Triple diagonal modeling is one tool to help map the journey the organization will take. It gives a view of the forest, not the trees, but has enough detail to help clarify the key performance problems and integration issues the organization must overcome.

Acknowledgment

This chapter is an expanded version of L. O. Levine and L. D. Villareal, "Triple Diagonal Modeling: A Mechanism to Focus Productivity Improvement for Business Scenarios," pages 406–11 in *Second International Symposium on Productivity and Quality with a Focus on Government*. Published with the permission of the Institute of Industrial Engineers Press, copyright ©1993.

References

Blackstone, J. H. 1989. *Capacity management*. Cincinnati, Ohio: South-Western.

Camp, R. C. 1989. *Benchmarking: The search for industry best practices that lead to superior performance*. Milwaukee, Wis.: ASQC Quality Press.

Davenport, T. H. 1993. *Process innovation*. Cambridge, Mass.: Harvard Business School Press.

Denton, D. K. 1990. *The production game: A user's guide*. Reading, Mass.: Addison-Wesley.

Forrester, J. W. 1992. Policies, decisions, and information sources for modeling. *European Journal of Operational Research* 59 (1): 42–63.

Gabriel, T., J. Bicheno, and J. E. Galletly. 1991. JIT manufacturing simulation. *Industrial Management & Data Systems* 91 (4): 3–7.

Hammer, M., and J. Campy. 1993. *Reengineering the corporation: A manifesto for business revolution*. New York: Harper Collins.

Isaacs, W., and P. Senge. 1992. Overcoming limits to learning in computer-based learning environments. *European Journal of Operational Research* 59 (1): 183–96.

Levine, L. O., and L. D. Villareal. 1993. Triple diagonal modeling: A mechanism to focus productivity improvement for business scenarios. In *Proceedings of the second international symposium on productivity and quality improvement with a focus on government*. Norcross, Ga.: IIE Press.

Marca, D., and C. L. McGowen. 1988. *SADT: Structure analysis and design techniques.* New York: McGraw-Hill.

Plossl, G. W. 1991. *Managing in the new world of manufacturing: How companies can improve operations to compete globally.* Englewood Cliffs, N.J.: Prentice Hall.

Senge, P. M. 1990. *The fifth discipline: The art and practice of the learning organization.* New York: Doubleday/Currency.

Senge, P. M., and J. D. Sterman. 1992. Systems thinking and organizational learning: Acting locally and thinking globally in the organization of the future. *European Journal of Operational Research* 59 (1): 137–50.

Shunk, D., B. Sullivan, and J. Cahill. 1986. Making the most of IDEF modeling—The triple-diagonal concept. *CIM Review* 3 (1): 12–17.

Stalk, G., P. Evans, and L. E. Shulman. 1992. Competing on capabilities: The new rules of corporate strategy. *Harvard Business Review* 70 (2): 57–69.

Suri, R. 1988. A new perspective on manufacturing systems analysis. In *Design and analysis of integrated manufacturing systems,* edited by W. D. Compton. Washington D.C.: National Academy Press.

Watson, G. H. 1992. *The benchmarking workbook: Adapting best practices for performance improvement.* Cambridge, Mass.: Productivity Press.

Wu, N. L. 1989. Understanding production systems through human simulation: Experiencing JIC (just-in-case), JIT (just-in-time), and OPT (optimized-production technology) production systems. *International Journal of Operations and Production Management* 9 (1): 27–34.

CHAPTER 4

Integrated Work Process Redesign
Monty L. Carson

Once management has begun creating and disseminating a vision for an agile corporate structure and culture, considerations can be made for redesigning the organization's work processes. Work process changes are where the real changes in an organization get made—not only the physical changes in product flows and plant layout, but in people's minds and attitudes. As people redesign their work processes, they will have to change their behavior, which leads to an internalization of agility in individuals and an institutionalization of agility in the organization.

For most organizations, there are many work processes that need to be redesigned. Redesign of an organization's work processes must be *sequential*, because they cannot all be redesigned at once. Redesign of an organization's work processes must be *ordered*, because some processes should be redesigned before others. Initial attention should be given to the shop floor, a manufacturing organization's primary value chain, where material is transformed and wealth created. As shop floor processes are redesigned, the administrative and support processes that are impeding agility will become apparent. These are the processes that should be redesigned next.

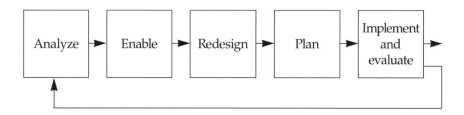

Figure 4.1. The integrated work process redesign (IWPR) framework.

This chapter will present a broad framework for conducting and managing redesign efforts in both shop floor and administrative environments. This framework, called integrated work process redesign (IWPR), consists of five general phases (see Figure 4.1). While every effort is unique and requires tailoring, the five phases of IWPR describe the essential elements necessary in a successful redesign effort.

This chapter begins with a brief historical review of some of the key principles of work redesign. Next, the IWPR framework and its prerequisites are discussed. This section is followed by a discussion of the training and tools that may be needed during IWPR. Finally, the pitfalls often encountered in improvement efforts are given with suggested remedies.

A Historical Review of Work Redesign Principles

The notion of studying work processes with the intent of improving them was born in the work of Frederick Taylor and his promotion of "scientific management" (see, for example, Taylor 1915). Taylor and those who followed him developed a set of theories and tools addressing the optimization of work, primarily at the worker level. This led to a whole new breed of professionals—industrial engineers—whose intent was to rigorously document and optimize work. Most recent literature and practice has denounced the idea of an expert being the only one capable of optimizing work and work processes. While the appropriate level of involvement of work experts is debated even today, Taylor's underlying message about productivity is undeniable: Great improvements are possible with a structured approach.

W. Edwards Deming shed a new light on understanding work processes when he voiced concerns over the popular conception that quality could be improved and costs lowered by pressing the workers to do better. He saw through the eyes of a statistician that defects and poor productivity were more the result of faulty work systems than of inconsiderate work by individuals. His message to those who would listen: The system must be improved. He urged management to take responsibility for improving work systems and to equip people for doing so through education. Deming promoted educating every employee to understand systems through statistics and letting them use that knowledge right away by improving their work processes, as people lose what they learn if they don't use it. However, he provided few clues on how to systematically approach getting people involved. Whatever he lacked in explicit instruction, Deming provided in an important and clear vision—empower the workforce through education, training, and job redesign.

Others have continued to promote involving the people in the work process in the process design. Gerald Nadler (1981) promoted a design approach that uses all the people involved in the process (not just "experts") in designing innovative work systems. In his approach, he sought to overcome lack of design expertise with a set of exercises that encourage people to develop a constructive mind-set and thoroughly explore the problem before starting to develop solutions. He recognized that optimal designs could only happen when designers were given a design structure and tools.

Marvin Weisbord (1987) suggested an approach for getting people involved in redesigning their workplace in which management provides direction and energy, while workers contribute critical inquiry and on-the-job experience to redesigning a better work environment. He demonstrated that management direction and employee design were both necessary—a work redesign without one or the other was seriously hampered.

Even when organizations are fully committed to employee involvement, they must be careful to avoid focusing on activity rather than results. In their analysis of recent improvement efforts, Schaffer and Thomson (1992) found that efforts that focus primarily on employee involvement are likely to fail. In other words, organizations that take an activity-centered approach focus on the amount of effort expended to produce improvement, measured in tangibles like the number of teams formed and trained. They know how much time and money they've spent getting employees involved, but are not able to articulate the result. Conversely, organizations that take a results-centered approach encourage specific innovations to meet measurable goals via short-term, incremental projects. This keeps motivation high, and leads to higher rates of success. The lesson here is that employees must be given concrete goals when asked to redesign their work processes.

The history of work redesign has produced a body of knowledge that has only been touched on here. We can extract from this knowledge a set of principles to guide redesign efforts. In summary, successful redesign efforts

- Have a structured approach
- Focus on the process and applying system principles
- Provide structure, tools, and skills to those redesigning work processes
- Have clear management direction, but primarily depend on those involved in the process
- Seek to accomplish specific, measurable goals

Integrated Work Process Redesign

Setting the Stage

IWPR cannot happen without a good team. Successful team performance requires careful team development. An excellent book on creating high performance teams is *The Wisdom of Teams* (Katzenbach and Smith 1994). Katzenbach and Smith define the key elements of successful team performance as (1) a small number of members, (2) complementary skills, (3) a common understanding of the team's purpose and a common commitment to challenging performance goals, and (4) a common approach to working together.

It is important for IWPR teams to have a relatively small number of members so that all members have an opportunity and an obligation to contribute to the team's work. When teams become too large, members feel less vital and may not perceive the opportunity to contribute. The phenomenon known as social loafing occurs when a member of a group feels his or her contribution is not identifiable. Responsibility for performance can become diffuse in large groups. Everyone assumes someone will do a particular task, and no one does. Thus, effective teams usually number anywhere from three to 20 members, but rarely more. This does not mean there won't be a large group of people contributing to the IWPR team's work, but the core team should be kept small.

IWPR team members must also bring complementary skills to the team. Important skills fall into three categories: (1) problem-solving skills, (2) interpersonal skills, and (3) technical skills. Problem-solving skills can be developed with redesign training, discussed later in this chapter. Interpersonal skills are very important and include communication, conflict management, active listening, and other "soft" skills. A team facilitator with good interpersonal skills plays an important role in helping team members continue to develop their interpersonal skills throughout the IWPR process. The role of the facilitator is discussed later in this chapter. Technical skills include general knowledge about making processes more agile and specific knowledge on the process that is being redesigned. The person who has general knowledge (referred to as the *guide*) and can teach it to the whole team (during enabler training) plays an important role in IWPR. The roles of the technical guide and enabler training are discussed later in this chapter.

To get individuals with specific knowledge of the process being redesigned, the IWPR team is initially formed with representatives from the target work process, representatives from processes that feed (upstream) or are fed by (downstream) the process, as well as support functions that play an important role in the process. There are no hard-and-fast rules about initial team formulation, except that everyone gets a voice. It is important to remember that insights gained after the process is mapped and analyzed may indicate that others need to be involved. Thus, the team may be reformulated based on a clearer understanding of the issues involved.

The team's common purpose comes from the specific goals that have been set in the organization's system-level analysis described in the previous chapter. The system-level analysis should be complete enough that the goal can be quantified with a performance measure. For example, a good goal for a team might be to redesign a shop floor process to reduce the cycle time of a product family by 50 percent or to reduce defects to 1 percent. The goal(s) should be attainable but challenging. It should be noted that goals can be counterproductive if misused. Easy goals fail to challenge teams to rethink the process, while overambitious goals frustrate and demoralize teams. In addition, team performance measures need to be designed to encourage teams to do their best, not to reward or punish merely on the team's success in meeting the initial goal. Realistic goals can only be developed when the process and its impact on the entire manufacturing system is understood, and when both managers and team members are involved in their development. The system-level analysis described in chapter 3 and the involvement of both management and workers as described throughout this book should result in the development of appropriate goals.

As mentioned in the chapter introduction, redesign efforts should start on the shop floor, where value is added to the organization's products. As the production processes on the shop floor are redesigned for agility, it will become apparent that many of the processes that support the shop floor need to be redesigned also. For example, during the redesign of an engine assembly line, it may become clear that production time is slow because the right parts never seem to get to the shop floor on time. This would point to a need to redesign production planning and control. Thus, a recognized need spawns goals for support processes, and new IWPR efforts can be started. The organization's change effort starts with its primary processes and lets the redesign of those processes reveal which other processes are important. The work process redesign efforts are integrated because they are identified and prioritized by their impact on shop floor agility.

The team's commitment to its purpose must be developed. It involves giving the team legitimacy and a structure, and preparing it for accomplishing its purpose. The key activities in preparing the team may be summarized as follows:

- Legitimize and solidify team.
- Establish formal roles on the team.
- Explore personality types and informal roles.
- Determine the team's operating mode.

The team is *legitimized* when management officially recognizes its purpose. The team members must also recognize the need for accomplishing the purpose and agree that accomplishing the purpose is worth effort and sacrifice for both the organization and employees. The team will begin to *solidify* when it understands its role and begins to get a sense of its own competence and value. The team needs to realize that teamwork is built on trust, honesty, and commitment—things that are developed over time. The team needs to be encouraged to enjoy the times when things are going well and to persevere when things are tough. There are a variety of team-building exercises, ranging from simple games to elaborate simulations, that can effectively be used at this time. The Association for Quality and Participation's resource catalog (see the "Sources of Additional Information" section in this chapter) contains a number of books on teams, team facilitation, and team building. In addition, see Wellins, Byham, and Wilson (1991) and Byham (1988) for more information on developing teams.

To help the team function effectively, several formal roles need to be established. Generally, three key roles must be filled; one person may fill more than one role. The first role is the *team facilitator*. The person in this role guides the team through the IWPR framework, keeps the team focused on the goal, works with the team on cultivating interpersonal skills, and acts as an intermediary to management. The team facilitator may be a member of the organization who specializes in facilitation, an internal person who has gone through the IWPR process before, or an external person who is consulting for the team. The second role is the *team administrator*. The administrator performs clerical duties such as scheduling meetings, keeping meeting notes, and tracking the results of the team's work. The administrator role may be shared by various team members over time. The third role is the *guide*. The guide is a technically oriented person who understands the principles of agility on the shop floor and throughout the organization (for example, an industrial engineer), and is responsible for introducing these principles in the enable phase of IWPR. The guide can also help direct the team as it redesigns its work processes. Because of the specific nature of the role, the guide doesn't need to be at all of the team meetings. In fact, the team will function better if the guide is not always available, because the team must work through issues on its own.

Part of the facilitator's role is to help the team understand and explore the way that each person's style and strengths contribute to the effectiveness of the group. People can contribute to the team in many different ways, as shown in Table 4.1. These differences are one of the strengths of the team concept, as they allow different people to fill different roles during the redesign effort. For example, some parts of the redesign effort require "bouncers"—people with creativity, new ideas, and energy. Other parts of the redesign effort require "plodders"—people who methodically seek data and scrutinize every detail. A team made up of only bouncers or only plodders will not succeed—both are needed. The team needs to actually discuss the preferred roles of each of the members and seek to find appropriate tasks for them. A number of questionnaires are available commercially that can be used to help identify team members' styles. Perhaps the most popular is the Myers–Briggs inventory (see Kroeger and Thuesen 1988).

Table 4.1. How people differ when working in teams.

Some people	While others
Think creatively	Think logically
Seek data	Make decisions intuitively
Think and communicate textually	Think and communicate graphically
Like to debate issues	Seek peace and consensus
Think out loud	Ponder quietly

Another useful questionnaire is the Kolbe conative index (Kolbe 1991). Either of these instruments will provide team members with valuable insights into the style differences of team members.

Finally, the team should discuss and develop a set of operating guidelines. These guidelines should explicitly state how the team will operate and how decisions are made. An excellent source for information and tools regarding operating guidelines for teams is *The Team Handbook* (Scholtes 1988). Some of the elements of the guidelines might be

- A group-generated list of ground rules that is posted in the meeting place. Example rules: criticize ideas, not people; all ideas are worth consideration; listen carefully to each other; don't dominate the discussion.

- A written plan describing how information will be shared between team members (for example, meeting schedules) and management (for example, progress briefings).

- A written agreement stating team member responsibilities, such as showing up to meetings on time and ready to participate, doing reading or homework assigned by the guide, and protecting the integrity of the team and other team members when talking with people who are not on the team.

Now that the team has been formed and prepared, IWPR is ready to begin. IWPR consists of five phases—analyze, enable, redesign, plan, and implement. These phases are explained in the next five sections, and the tools and techniques for each phase are discussed in the "Training" section later in this chapter.

Analyze

The purpose of the analyze phase of IWPR is for the team to discover and document the details of the work process and find what is hindering the process from meeting its target goal(s). At the end of the analyze phase, the team should have documented the current work process, measured the process in terms of the performance goal, and identified the problems with the process in its current condition.

The team needs to create an accurate picture of the work process in its current state, documented in a format that everyone understands. Team members that have been involved in the organizational-level analysis (described in chapter 3) will have been exposed already to the general production system. However, seeing the process in its entirety and in detail may be new to them, as it probably will be for most of the other team members. Several process documentation tools are discussed under "Training." Regardless of the process documentation tool used, the team should clearly identify several elements of the process. These elements include the process phases, process boundaries, and inputs and outputs (see Figure 4.2).

The entire process documentation can be done by the team, where all team members participate in a series of meetings to lay out the process. This is especially efficient if the participants already have a basic understanding of the process. The alternative is for a facilitator to capture the basics of the process through individual interviews and to prepare an initial version of the documentation for the team to review and complete. This can be more efficient when team members are familiar with only a subset of the work process. This first draft gets everyone to a common starting ground, from which they can finalize the current version and identify process problems.

An accurate picture of the current system is needed. This includes not only a representation of the process flow, but some data to measure the performance of the process in terms of the process goal(s). For

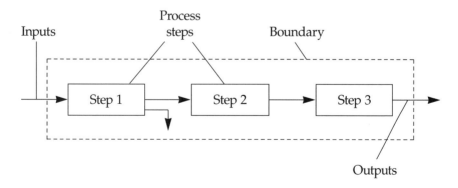

Figure 4.2. Elements that should be captured while documenting a process.

example, if the goal is to reduce defective parts to 1 percent, the team would collect data on the rejected product at each process step. Some other examples of data to be collected include the average amount of time a part waits for deburring, the amount of time spent reworking parts from suppliers, or the time it takes to get a work order to the floor. Because one of the key characteristics of agility is a short cycle time, the current cycle time for the process should always be collected.

Based on the data collected, as well as team members' understanding of the process, the problems with the current process can be identified. Problems should help explain why the process performs as it does. For example, long cycle times may be caused by too many products being on the shop floor at once, long setup times, lack of parts, and long transportation time between process steps. Defective parts may be caused by machinery that goes out of tolerance, untrained workers, or inconsistent work methods. Many of the problems will be familiar to the team members. This activity substantiates what they already suspect. The "Redesign Training" section gives many tools to help the team determine and document causes of process problems.

The analysis phase of IWPR is an iterative one, involving several cycles of collecting data and refining the process documentation. Throughout process analysis, team members will have ideas and opinions on how to improve the system. These should be captured without prejudgment for later use. When finished with the analysis phase of IWPR, the team should have a process diagram, the performance of the process in terms of the process goals, and an unabridged list of problems that hinder the work process.

Enable
The purpose of the enable phase of IWPR is to give the team the understanding of the agile practices that will help them redesign the work process and achieve the process goal(s). After analyzing the process, the team members have been stretched a bit. They now have a picture of the process in which they work, with all its bumps and warts. They have a goal to reach, and may or may not have any ideas on how to get there. The team members probably have a mixture of emotions: fear of the unknown, resistance to continue, fear and doubt about being able to

reach the goal, and curiosity about a better way to do things. There may even be a sense of crisis. Because people usually welcome new knowledge when they understand why they need it, this is the time to train the team in agility enablers.

Agility enabler training teaches the team about agile manufacturing practices and how those practices can be adopted and adapted to transform their own work process. Agility enabler training is meant to encourage, inspire, and stimulate. The team members learn that they are not the first to tread the difficult path of change. They learn that others have faced this challenge and successfully applied practices to their production processes to become agile manufacturers. Agility enabler training is a critical part of IWPR and is discussed later in this chapter.

The enable phase of IWPR should result in a team that has new knowledge about agile practices and has ideas on how to improve the work process. The best way to proceed once the team members have been trained in an agile practice is to immediately get them into redesigning their work process, realizing that there will be several iterations between the enable and redesign phases.

Redesign

The purpose of the redesign phase of IWPR is for the team to use its newly acquired knowledge on agile practices to redesign the work process to meet the process goal(s). The redesign phase of IWPR is where things start to get fun. The team, aware of the entire process, its problems, and the goals it needs to meet, has been trained on agile manufacturing practices that will help it redesign its process. Usually the team is anxious to start redesigning its work process. In fact, while getting the training in agile practices, team members have probably already started identifying ways to improve parts of the current process. In this phase, the team concentrates on the entire work process, redesigning it to meet the process goal(s) using agile practices. The group members have to learn how to create and design together. The "Redesign Training" section lists some of the tools and techniques the team can use to generate and document new design ideas.

There are several points that the team should keep in mind during this phase. First, there is no optimum solution to any manufacturing system problem. There will likely be several good ways to do any task. No design is ever perfect; it must be continually reevaluated and improved upon. This attitude will move the team far along into a mode of continuous improvement after formal redesign efforts are finished.

Next, improvements that are low cost and can be quickly implemented are "quick wins" and should be implemented immediately. The process doesn't have to be changed all at once. Small improvements demonstrate concepts, keep the team on the right track, and keep everyone's spirits high.

Finally, the team members are doing something they have probably never done before—making significant decisions about their workplace. Members should realize that after the redesign is done, they will have much more influence over the workplace. An agile manufacturing process requires agile people. The expanded influence of a worker in an agile environment may well represent a new mind-set for team members, as illustrated in Figure 4.3.

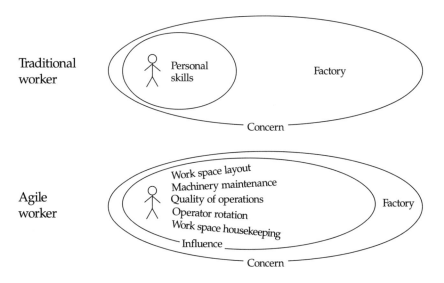

Figure 4.3. Traditional and agile worker's areas of influence and concern.

At the end of the redesign phase, the team should have tested and implemented the quick wins. In addition, it should have a clear understanding of the remaining changes that it wants to make in the work process.

Planning

During the planning phase, the team constructs tentative plans for implementing its redesign, presents its design for the new work process to management for approval, and finalizes plans for the implementation.

The team makes tentative plans for implementation using the tools described in the "Redesign Training" section. The team needs to be sensitized to alternative implementation strategies such as phased or instant switch over. The team also has to convince itself that the design is reasonable and can be implemented in a reasonable time frame.

The team must then present the design and implementation to management for approval. Training for presentations will probably be required, as many of the team members may never have been in this position before. The impressive part of an IWPR team presentation is that the team members are behind it, and they will sell it. The guide and/or any other outside team members are in the background for guidance and support. The team must convince management that the design has merit (that is, it will meet the process goal) and that the implementation is feasible. The effects of the transition on production should be carefully considered. The team must determine and clearly explain who else will be involved in the implementation and what their role will be. If all of the workers in the process have not been involved in the redesign, management may need to allocate resources for training. It is also possible that changes to other processes will be required to support this new process. If so, management will need to approve and allocate resources for another IWPR team. Depending on the extent and expense of the design, the team may have to request capital expenditures and need to show payback. Thus, cost, schedule, resources required, and benefits must all be presented. Once the design and implementation are approved, the team must finalize the design and implementation plans. The new design will probably require new procedures and work methods, for which documentation must be written. If training for the new work process is required, the team must plan and organize it. One of the beauties of IWPR is that the team members understand the new process intimately, are qualified to be trainers, and should be used as such.

At the end of the planning phase, the team should have clearly presented its design and implementation plan to management. Management can either approve the resources required to implement the new work process or provide direction that the team can use to modify its design and/or plans.

Implementation and Evaluation

In the final phase, the team leads the implementation of the new work process. This can include training other workers in the process, coordinating equipment moves, and tweaking the process to work out the problems. The data collection done by the team during analysis can be formalized to provide ongoing indicators of the process performance. By now, the guide is nearly out of the picture, the team is motivated, and there is no need to convince the team members what to do—they are the pushers. Management facilitates the resolution of unforeseen problems that arise during implementation. Data are collected on the process to see if the changes are having the desired effect. The team may need to tweak the current design using the agility enabler principles they already know, or move on to the next agility enabler. It may take several iterations through the IWPR framework for a team to completely implement all of the agility enablers that are possible. The "Training" section gives a recommended progression of agility enablers to implement.

IWPR projects should have a definite end. When the IWPR team has accomplished its purpose in redesigning the process, the team should be encouraged to keep reexamining and improving it, thus entering a state of continuous improvement. This may require the identification of a team member as the continuing facilitator. This facilitator would be responsible for keeping the documentation of the process current, monitoring (along with management and the team) the performance of the process, being the receiver of improvement ideas, and convening the team periodically to revisit the process design.

Team members are very valuable to the organization, for they have both received training and participated in IWPR. In fact, the team members are well qualified to assist in other similar IWPR projects. They can be especially effective as trainers for other IWPR teams with similar problems, because they understand the organization as well as the concepts they are teaching. However, new teams in other parts of the organization will still require a guide.

At the end of this phase, the team should have implemented a stable work process and be able to measure its performance. The team may need to return to the enable phase for another cycle of redesign, or it may move into a mode of continuous monitoring and improvement.

Training

IWPR is centered around solid systems and industrial engineering principles, yet it is dependent on the people in the process, not specialists, to be successful. Because IWPR teams are composed of individuals who probably have little experience designing processes, training becomes a critical part of IWPR. To successfully redesign the process, team members need to understand the principles, methodologies, and tools of process redesign and the principles and practices that enable agility in manufacturing. Training in process redesign will allow the team to document, measure, evaluate, and communicate work process changes. The team also needs to be trained in agility enablers. These are the principles and practices that will allow the team to redesign the existing process into an agile one. As illustrated in Table 4.2 there are subtle but important differences between training in redesign and training in agility enablers. The next two sections discuss each type of training in detail.

Redesign Training

The IWPR team needs to be trained in redesign principles in every phase of IWPR. This training should include a review of the purpose of the phase, the activities that will happen during the phase, and the desired outcome at the end of the phase. In addition, the team needs to be trained in the tools that will help during each phase. This section lists some of the training and tools that may be needed by the team

Table 4.2. Differences between redesign training and agility enabler training.

Process redesign training	Agility enabler training
Teaches analytical tools and techniques applicable to many types of processes	Teaches agile manufacturing practices that apply to only certain types of processes
Gives the IWPR team the tools to redesign the process	Gives the IWPR team the insights to redesign the process
Is provided in each phase of IWPR	Is provided during the enable phase of IWPR
Examples	Examples
—Process documentation	—Cellular layout
—Idea generation	—Setup reduction
—Project planning	—Preventive maintenance

during each of the IWPR phases. Most of these tools are explained in other publications and are only referenced here. One tool, user-centered diagrams, is explained following this section. Not every tool is needed every time; one of the roles of the guide is to determine which tools are appropriate for the team's situation.

Tables 4.3–4.6 show the training that needs to be provided for each IWPR phase, and some of the tools that could be used. References for further information are included in these tables.

As shown in Table 4.3, a user-centered diagram (UCD) is a tool that is useful in some redesign efforts. Because it is not well documented in other literature, a brief description is included here.

User-Centered Diagrams—A Redesign Tool

The appropriate tool or tools for documenting the process and redesign depends on the characteristics of the process and the scope of the IWPR effort. For example, for redesign efforts of small scope (for example, material processing on a limited part of the shop floor or an administrative process in a single department) a simple process flowchart and a work space layout will often be sufficient.

Table 4.3. Training and tools for the analyze phase.

Training	Tool	Description
Process documentation	UCD	Graphical depiction of process steps using icons[1]
	Process flowchart	Graphical depiction of process using standard symbols[2]
	Work space layout diagram	Top view of equipment location and material flow[2]
Data collection	Check sheet	Data collection device[3]
	Traveler	Paper attached to an item traveling through the system to record time and other data
Problem identification		
—Identifying variation	Histogram	Bar chart showing the frequency of occurrences[3]
	Run chart	Sequential plot of data values to show trends[3]
	Control chart	Sequential plot of data values to show process stability[3]
—Identifying root causes	Fishbone diagram	Chart showing sources of defects: categories include people, equipment, materials, methods, and environment[3]
	Scatter diagram	Data plot to show the potential relationship between possible cause and effect[3]
	Pareto diagram	Bar chart of causes to determine significance[3]

1. Discussed later in this section.
2. Ishiwata 1991.
3. Cooksey, Beans, and Eshelman 1993.

Table 4.4. Training and tools for the redesign phase.

Training	Tool	Description
Idea generation	Brainstorming/ Blue sky	Activity to generate creative redesign ideas.[1]
	Colored hat thinking	Allows ideas to be generated and evaluated in a number of thinking modes.[2]
	Challenge assumptions	Activity to examine and overcome limiting assumptions.[1]
	Solution map	Activity to generate redesign ideas.[1]
Idea evaluation	Force field analysis	Examination of pros and cons of potential redesign idea.[1]
	Cost/benefit analysis	Trade-off study of the costs of an idea and the returned benefit.
	Risk analysis	Activity to categorize risks in terms of probability and consequence.
Process redesign	Process documentation	See tools in Table 4.3.
	Work space chalk-up	Use chalk to draw proposed layout in a parking lot at full scale.
	Miniature lab simulation	Simulate proposed work flow on paper or in miniature to visualize and test it.
	Computer simulation	Simulate the proposed work process with a software package to predict performance.

1. Cooksey, Beans, and Eshelman 1993.
2. Buzan 1982.

Table 4.5. Training and tools for the plan phase.

Training	Tool	Description
Project planning	Action plan	Lists actions, responsible person, completion date, resources required, and so on[1]
	Gantt chart	Shows activities and milestones in terms of time[1]
	Pert chart	Shows activities and their required order of completion
Presentation making	Overhead view graphs	Presents concise statements of ideas being presented
	Presentation package	Presents photocopies of view graphs, charts, diagrams, and other information

1. Cooksey, Beans, and Eshelman 1993.

Table 4.6. Training and tools for the implement and evaluate phase.

Training	Tool	Description
Project management	Action plan	Tracks actions and responsibilities (see Table 4.5)
	Gantt chart	Tracks actual progress against schedule
Communication to others	Bulletin boards	Displays implementation status and process performance
	Flyers/ newsletters	Notifies others of changes, keeps others up to date
	Tours	Exhibits and explains new work process to others
Evaluation	See tools in Table 4.3	

However, work redesign efforts often involve (a) interactions between shop floor and support functions, (b) information flows in multiple formats (paper forms to keyboard entry to electronic transmission), and (c) differences in processing schemes (process on arrival versus process in batch). A tool that is often appropriate for these work redesign situations is called user-centered diagramming. A UCD is a tool that combines graphics and text to describe a work process and detail human tasks (see Figure 4.4). The flow of the workpiece is traced through the steps of the process along the horizontal axis. Process steps are associated with function or physical locations along the vertical axis and time along the horizontal axis. Process steps are represented as icons (manual operations, machines, computers, in boxes), rather than the shapes of traditional flowcharting (boxes, circles, triangles). Data (time, rejection rates, and so on) can be represented next to the associated task.

A representation of the process that everyone on the team can see in its entirety is desirable. Thus, UCDs are most effective when they are large, covering an entire wall. UCDs can be developed manually— yellow sticky notes on a white board (or large piece of paper) works best for the initial development. If multiple changes to the UCD will be made, the UCD can be transferred to a graphics software program. Computers with scalable printing capabilities allow the UCDs to be enlarged and reprinted as the documents are updated. Large UCDs give plenty of room for team members to record descriptions, exceptions, comments, and suspected process hindrances.

Agility Enabler Training

Most process improvement approaches are based on the premise that the people involved with the process have the greatest potential for improving the process and making the improvement work. That makes sense. However, many of these approaches also assume that participants already have the knowledge they need to come up with the best solution. Thus, team training usually focuses on the tools discussed in the previous section. This is not the surest way to optimal solutions for several reasons.

First, there are many practical, proven ideas for increasing agility in a work process. Team members who have been exposed to proven

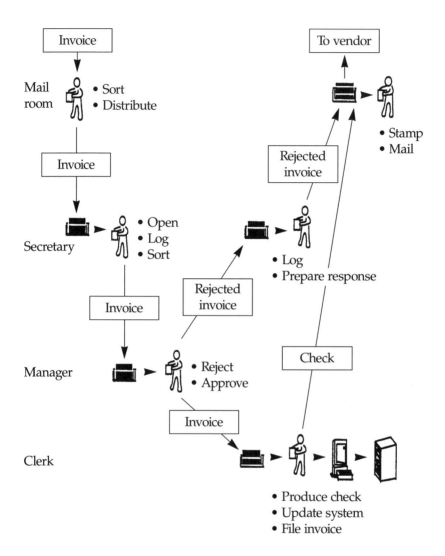

Figure 4.4. A simple accounts payable process to illustrate user-centered diagrams.

agile practices will have a broader range of potential solutions to problems they have encountered. This is called expanding the solution space. Second, few people think about work processes as a system. That is, they don't think about the principles that govern systems in general and how those apply to a specific work process. These principles deal with issues such as queues and how they are managed, information and how it is communicated, and flexibility and how it is achieved. In manufacturing, these principles have been discovered over time and distilled into the agile practices discussed in chapter 5. Understanding why these agile practices work increases a team member's ability to correctly apply that practice to his or her work process. Thus, team members who are trained in agile practices should be taught two things: *how* agile practices work (the application on the shop floor) and *why* (the underlying principle).

Knowledge inspires creativity. It is difficult for people to create new ways of doing things from scratch. In addition, people tend to be limited in their thinking by self-imposed constraints of which they may not even be aware. But when they have been exposed to a new mindset, people tend to lose those constraints and can visualize doing things they've never done before. The concept of setup reduction is a good example of this. For years, long machine setup times were accepted as an unchangeable part of manufacturing. Production was scheduled in large batches to minimize the effect of setup. Huge amounts of effort were focused on reducing the effects of long setup time, rather than reducing the setup time itself. But a series of realizations by Shigeo Shingo, a Japanese engineer, led to the understanding that setup time could be reduced through the application of some simple, but powerful, concepts. Shingo cites case after case where machinists and line supervisors, previously convinced that long setup time was unavoidable, reduced setup times from hours to minutes using setup time reduction. These people merely needed to have a faulty belief challenged and to be given a new way to look at the problem.

Introducing agility to a manufacturing process usually involves the infusion of a number of agile practices. There is a natural progression to the introduction of those practices, based on the concepts of value chains. A value chain is the series of activities in a process that truly increase the value of the final product. In manufacturing, a good portion of the value chain exists on the shop floor, where raw materials are converted to valuable end products. Other functions exist to support the activities on the shop floor. When introducing agility to a manufacturing organization, a sound strategy is to start at the shop floor and work toward support processes. This is the strategy that Black (1991) proposes. Figure 4.5 illustrates this concept, showing many of the processes that make up a manufacturing production system. All processes exist to support the production process shown at the bottom of the figure. In Figure 4.5, boxes represent processes that the team can affect directly while redesigning a shop floor process. Ovals are processes that support and interact with the shop floor process, but would generally need a separate IWPR effort to redesign. As illustrated, processes further from the shop floor have a diminishing affect on shop floor agility.

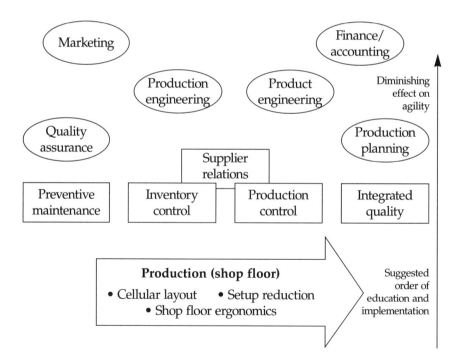

Figure 4.5. Relation of processes in a manufacturing production system.

Following the concept illustrated in Figure 4.5, a suggested order of introducing agility into a shop floor redesign is given in Table 4.7. Notice that the second agility enabler is "Shop floor ergonomics," which should include consideration of safe lifting techniques; providing mechanical aids to minimize lifting heavy loads; making workstations adjustable so that workers of different sizes can avoid awkward stationary, working, and lifting postures; and avoiding repetitive motion injuries. Besides the obvious benefit of protecting the individual worker, application of shop floor ergonomics concepts usually makes the workplace more flexible, a critical component of agile manufacturing. Training on shop floor ergonomics will allow the team to redesign its work process with these concepts in mind.

Inexpensive sources of prepared educational material in these areas are not easy to find. Table 4.8 lists widely available sources of information from which educational material can be prepared. Many of the sources are books (listed by author) and catalogs that are contained in the "References" and "Sources of Additional Information" sections at the end of this chapter.

The format and delivery of the training material to the team will depend on the size and background of team. I have found that two or three classroom sessions spread over several days works well. This gives the team members time to think through the concepts being presented and to apply the concepts to their current work environment. There is no better way to start a training session than with the testimonial of an excited team member who used a new concept to the betterment of his or her current job. Some agile manufacturing practices, such as setup reduction, lend themselves to immediate application. Others, such as supplier integration, require systematic changes to the work process to implement. Such systematic changes will require formation of new teams to address such issues.

Another very useful method of training a team is through benchmarking, which is investigating the processes of other organizations that have been recognized for their superior performance, usually through on-site visits. After the team has documented and measured its current process and has been educated so that it understands the

Table 4.7. A suggested order of introducing agility on the shop floor.

Agility enabler	Effect on agility
1. Cellular layout	Putting process steps close together on the shop floor minimizes material handling costs, eliminates non–value added time, increases communication (information flow), and increases flexibility of workers to move between tasks.
2. Shop floor ergonomics	Building flexibility into the workplace allows workers to efficiently share tasks while protecting them from injury.
3. Setup reduction	Setup reduction allows minimal batch sizes to reduce work-in-process between activities.
4. Integrated quality	This minimizes waste due to production of defective parts by defect prevention (foolproofing) and defect prediction (quality control).
5. Preventive maintenance	Preventive maintenance allows reduction of work-in-process inventory by eliminating need for buffer stock to cover for failed machinery.
6. Production control	Pull production using simple control techniques such as kanbans minimizes work-in-process inventory, cycle time, and overhead costs.
7. Inventory control	This works in conjunction with pull production to provide parts to the shop floor only when needed and to minimize inventory.
8. Supplier relations	This integrates suppliers into shop floor operations and features frequent delivery to the shop floor, standard order quantities, certification for quality, and single-source, long-term contracts.

Table 4.8. Sources of material for agility enabler training.

Agility enabler	Source
Cellular layout	Black 1991, chapter 4 Sekine 1992 Harmon and Peterson 1990, chapter 5
Shop floor ergonomics	National Safety Council 1988 National Safety Council 1993
Setup reduction	Shingo 1985 Black 1991, chapter 5 Harmon and Peterson 1990, chapter 7
Integrated quality	Black 1991, chapter 6 ASQC Quality Press catalog; books on quality control
Preventive maintenance	Productivity Press catalog, books on total preventive maintenance Black 1991, chapter 7
Production and inventory control	APICS educational catalog; books on production and inventory control Greif 1991 Black 1991; chapter 9
Supplier relations	Black 1991, chapter 10 Maass, Brown, and Bossert 1990

agile principles that can be applied to the process, benchmarking can provide the team with working applications of those principles in other organizations. It should be noted that it is in this context that benchmarking is an effective learning mechanism. Unless the team understands the problems of its own process and the principles that will help improve it, benchmarking trips will be little more than industrial tourism.

Training in agility enablers is a key part of IWPR. Together with team-based work redesign, it is a powerful force for change. People who know their work process are encouraged, empowered, and educated; their potential for success is high.

Pitfalls to Avoid

In a redesign effort, there are a number of potential pitfalls that can ruin or seriously hamper the effort. This section contains a number of these potential pitfalls and suggests appropriate alternative actions.

• *Lack of time for the team.* The team must be together a significant amount of time to be productive in the IWPR process. This usually means at least one to two days a week.

• *Lack of commitment of resources for the team.* If top management is not interested in providing adequate time and budget for the redesign team, the redesign effort should be stopped. As discussed in the first part of this book, successful change efforts must address strategic issues in the organization and take adequate steps to overcome resistance. If the current redesign effort has not been put in this context, it will likely fail. A lack of commitment of resources is an indication that top management has not done the up-front work to ensure a successful redesign effort. Proceeding further is very risky.

• *Avoiding redesign efforts because of limited resources.* Many companies, especially small ones, struggle to find the resources to provide IWPR groups with facilitation, training, and adequate time. Potential sources of low-cost assistance such as state- and federally funded manufacturing resource centers, universities, and national laboratories should be investigated.

• *Getting unhelpful assistance from outside sources.* Having said that low-cost assistance is available, be aware of low-cost sources of assistance that push a specific agenda. For example, some universities can provide supervised graduate students at a low cost. However, universities are constantly under pressure to publish research. Professors and students may be interested in helping only if they can try something new and different. It is important to get assistance from people that understand agile practices and want to see these implemented correctly.

• *Team members not working together well.* If the redesign effort is in the early stages, team-building exercises can help people understand each other's differences and learn to work together. If the redesign effort is well underway, conflict resolution techniques might be needed. The Association for Quality and Participation publishes a resource catalog (see "Sources of Additional Information") that has several good resources to help team facilitators.

• *Resistance of midlevel management.* Agile manufacturing and the changes that it brings (for example, focused teams and employee involvement and empowerment) can be very threatening to the traditional manager. Unless the manager has been educated in agile manufacturing and has bought into the concept, the manager will likely subvert and sabotage efforts to become more agile. A redesign effort should not proceed if midlevel management is not supportive.

• *Unenlightened managers participating on the IWPR team.* Supervisors, first line managers, and others in middle management can be a great asset to an IWPR team because of their knowledge of the shop floor or their functional area. However, the manager who does not understand and actively promote his or her appropriate role during the IWPR activity can hinder the team in different ways. A manager who takes charge will dampen much of the interaction that allows the group to form an effective team. A manager that avoids taking a leadership role in the group can hinder the group as well, as people have a tendency to defer to the person in highest authority. Therefore, an enlightened manager/team member recognizes that he or she is only a contributor, and works hard to make sure everyone else understands this.

• *Starting IWPR in the wrong area of the organization.* Almost any team that is assembled can find some streamlining opportunities within its immediate sphere of influence. However, very soon the team will be forced to interact with other parts of the organization to continue making improvements. The real art in implementing change in an organization is doing the individual improvement activities in the right order. In manufacturing organizations, it is usually best to start with improving the key manufacturing processes, then starting and linking improvement activities in functions that have key supporting roles to the shop floor.

• *Diving in too deep, too soon.* While key manufacturing processes should be improved first, it is often a mistake to tackle the biggest problem area first. An organization needs a project in a visible yet noncritical area to demonstrate IWPR and generate enthusiasm.

• *Adopting an inappropriate technology solution.* Technology such as automation is often seen as the ultimate in modernization. However, technology implementations are high risk and are only appropriate in certain situations (see chapter 6). For example, it is generally not profitable to automate a process unless there is high volume, low-variety production, or health and safety reasons to do so.

• *Adopting a technology solution too early.* Even in cases where technology is appropriate, the effect of the technology is diminished when the underlying process is inefficient. The adage of "simplify, automate, integrate" almost always holds true.

• *Lack of a tangible, specific goal.* An IWPR team must be given a definite destination so that it knows if it is headed in the right direction and when it has arrived. Without a well-stated goal that everyone understands, the redesign team can quickly lose focus and accomplish nothing. A specific goal is one of the prerequisites of a redesign team.

• *Overlooking small improvements.* Small, inexpensive, quickly implemented solutions should be encouraged. As Deming (1986, 127) said, any improvement is important, because it gives the participants a sense of pride, the value of which cannot be measured.

• *Implementing permanent solutions.* Solutions should be sought with a principle in mind—What is appropriate today may not be tomorrow. One of the traits of an agile organization is its ability to reconfigure its shop floor and support functions easily. Teams should be careful not to make changes that improve current operations, but will be difficult to change in the future. For example, conveyors almost always decrease flexibility, making the shop floor more difficult to operate and reconfigure.

• *Taking too long to go through the process.* The longer the period between when the redesign effort starts and when change is implemented, the harder it is to make significant improvements. People in the organization who don't support change can use this time to think of reasons why changes should not be made, as well as to undermine management support of the improvements. Once improvements are identified, approved, and planned, they should be implemented as soon as possible.

• *Waiting for the perfect plan when a good plan exists.* The IWPR team and management must tread a careful line between implementing without adequate planning and planning without ever implementing. There are several ways in which an organization's approach to improvement affect the team's ability to strike the right balance. First, an initial improvement effort in a visible but noncritical area will provide lessons learned and confidence for later teams. Second, teams should be assured that perfection is not expected once the implementation is complete and that repeated refinement of the implementation (continuous improvement) is expected. Finally, a focus on agility means that solutions should be simple and flexible, which lowers implementation cost and risk.

Conclusion

The heart of a manufacturing organization is its manufacturing processes. They accomplish its purpose and support its existence. However, changing those processes is neither easy nor painless. It requires a sustained, organization-wide change effort. Management and support processes must be made to align themselves with agile manufacturing changes on the shop floor. Success is far more likely if the reins of change are given to the people who rightfully own the processes. IWPR is a way of doing just that.

References

Black, J. T. 1991. *The design of the factory with a future.* New York: McGraw-Hill.

Buzan, T. 1982. *Six colored hats.* New York: Little Brown.

Byham, W. 1988. *Zapp! The lightning of empowerment.* New York: Ballantine Books.

Cooksey, C., R. Beans, and D. Eshelman. 1993. *Process improvement: A guide for teams.* Arlington, Va.: Coopers & Lybrand Publishing.

Deming, W. E. 1986. *Out of the crisis.* Cambridge, Mass.: MIT Center for Advanced Engineering Study.

Greif, M. 1991. *The visual factory.* Cambridge, Mass.: Productivity Press.

Harmon, R., and L. Peterson. 1990. *Reinventing the factory: Productivity breakthroughs in manufacturing today.* New York: Macmillan.

Ishiwata, J. 1991. *Productivity through process analysis.* Cambridge, Mass.: Productivity Press.

Katzenbach, J., and D. Smith. 1994. *The wisdom of teams: Creating the high performance organization.* New York: Harper Business.

Kolbe, K. 1991. *Kolbe conative index.* Phoenix, Ariz.: Kolbe Concepts.

Kroeger, O., and J. Thuesen. 1988. *Type talk.* New York: Delacorte Press.

Maass, R., J. Brown, and J. Bossert. 1990. *Supplier certification: A continuous improvement strategy.* Milwaukee, Wis.: ASQC Quality Press.

Nadler, G. 1981. *The planning and design approach.* New York: John Wiley & Sons.

National Safety Council. 1988. *Making the job easier: An ergonomics idea book.* Itasca, Ill.: National Safety Council.

National Safety Council. 1993. *Ergonomics: A practical guide.* Itasca, Ill.: National Safety Council.

Schaffer, R. H., and H. A. Thomson. 1992. Successful change programs begin with results. *Harvard Business Review* (January–February): 80–89.

Scholtes, P. 1988. *The team handbook.* Madison, Wis.: Joiner Associates.

Sekine, K. 1992. *One-piece flow: Cell design for transforming the production process.* Cambridge, Mass.: Productivity Press.

Shingo, S. 1985. *A revolution in manufacturing: The SMED system.* Cambridge, Mass.: Productivity Press.

Taylor, F. W. 1915. *The principles of scientific management.* New York: Harper and Row.

Weisbord, M. 1987. *Productive workplaces.* San Francisco: Jossey-Bass.

Wellins, R., W. Byham, and J. Wilson. 1991. *Empowered teams.* San Francisco: Jossey-Bass.

Sources of Additional Information

American Production and Inventory Control Society (APICS)
500 W. Annadale Rd.
Falls Church, VA 22046
800-444-2742

American Society for Quality Control (ASQC)
Quality Press
611 E. Wisconsin Ave.
Milwaukee, WI 53202
800-248-1946

Association for Quality and Participation (AQP)
801-B W. 8th St.
Cincinnati, OH 45203
513-381-1959

Institute of Industrial Engineers (IIE)
Industrial Engineering & Management Press
25 Technology Park/Atlanta
Norcross, GA 30092-2988
404-449-0460

National Safety Council
1121 Spring Lake Dr.
Itasca, IL 60143-3201
800-845-4672

Productivity Press
Box 13390
Portland, OR 97213-0390
800-394-6868

—

CHAPTER 5

Agile Practices
Cody J. Hostick

Small and midsize manufacturing firms of 500 employees or fewer owe their existence largely to the lack of responsiveness of larger manufacturing firms. The competitive advantages of smaller firms often consist of their ability to thrive in niche markets, involving significant amounts of custom work. This high-product-mix, low-volume-production environment is often viewed as unattractive to larger firms that lack the flexibility and responsiveness to compete. Smaller manufacturers can use the agile practices discussed in this chapter to help maintain and expand this competitive advantage. Since about 100,000 of the 360,000 manufacturing firms in the United States have 30 employees or fewer (McHose 1992), the agile practices discussed in the following sections have wide applicability to a significant percentage of the U.S. manufacturing base.

This chapter presents agile practices in the form of how-to checklists to help smaller firms using this information change the way they do business. Where appropriate, these checklists are augmented by lessons learned to assist firms in avoiding the mistakes of those that have gone before them. Agile practices are mapped to the product development loop and the manufacturing loop presented in Figure 5.1.

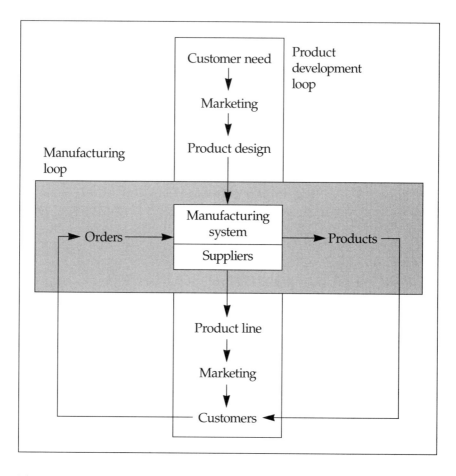

Figure 5.1. Manufacturing system.

The purpose of this mapping simply serves to sort agile practices into their two areas of focus.

1. Getting a new product developed

2. Getting a product made

Agile practices for product development begin with techniques for early identification of customer needs. These needs are then translated into a product design and development effort. The manufacturing loop begins with establishing a system to translate product designs into products to be shipped, followed by execution of the manufacturing processes. Agile practices to assist the product development are presented in the following section. Manufacturing agile practices are presented later in this chapter.

Agile Practices for Product Development

Smaller firms have the same need as larger firms to be agile in product development, but for different reasons. Smaller firms need to be agile in new product development because new products often represent significant portions of the firms' future income. This is due to many small firms surviving on niche markets, resulting in smaller volume, higher mix product lines with significant new/modified/tailored products. Slowly getting these new products to market may threaten the income stream that many firms need to survive. Smaller firms also need to be agile in new product development simply because they do not have the money to be inefficient. Larger firms need to be agile in product development to maintain market share. In addition, the move toward smaller volume batch production offers less opportunity to recover up-front product/process design and tooling costs, resulting in a strong need for cost-effective development systems (Maddux and Jain 1986, 1–15).

Key supporting agile practices that can be of assistance to smaller manufacturers in the product development loop include first understanding the current product development process and how long it takes. Next, early identification of new product needs can enable firms to maximize the time they have for development activities. Finally, the development activities themselves can be facilitated by using and institutionalizing a number of specific agile practices. These practices are presented in the following sections.

Understand the Product Development Process and How Long Things Take

A prerequisite to achieving product development agility is to understand for your own firm how the product development loop, shown in Figure 5.1, works and how long it takes. In many small firms, new products simply "happen." Significantly new products tend to originate from the idea of a single individual (usually the owner). This idea is fabricated into a prototype and eventually a product, using resources borrowed from the shop floor. Often, this research and development work is treated as "filler" work, meaning that employees work on the new product during lulls in regular production activities. Other types of new products tend to be modifications to existing products. Many of these variant new products originate from conversations with the customer base. The first copy of these variant products are put together and refined to eventually represent a new product line.

Regardless of the specific product development cycle in small firms, several general observations and low-cost improvement opportunities can be made. First, most small firms do not have a formal product development system. While documentation by itself does not lead to benefits resulting from agile practices, it can help direct agile practices to where they can provide the most benefit.

The key performance metric in the product development loop shown in Figure 5.1 is the total amount of time required for a customer need to be translated into a product delivered to the customer. By documenting the existing product development process and how long things take, a small firm can identify where the greatest opportunities for improvement exist and focus the use of selected agile practices. Smaller firms that document their product development cycle find basic personal computer (PC) tools more than adequate to capture the necessary information. As a start, a simple log of daily activity is often adequate to understand the product development process.

What the product development process documentation usually finds is a tremendous amount of dead time, or time when nothing is being done. While in many cases this dead time is due to smaller firms not having the financial resources to devote constant attention to new products being developed, often the dead time can be attributable to development work not being aggressively managed. Determining how long things take by documenting the development process is important in achieving the most from investments in agile practices. For example, the time gained by investing in computer-aided design (CAD)/computer-aided manufacturing (CAM) computer integration might be dwarfed by the time gained in having back shifts fabricate prototype designs as part of their regular work duties. Tying in all agile practice improvements to the total development time can help keep the benefit of the improvement in perspective. While CAD integration might eliminate 62 percent of computer numerical control programming time, documentation of the entire process might indicate that this only represents 8 percent of the product development cycle time and is not worth the significant investment required.

Another benefit that can be derived from documenting the product development process as a precursor to agile practice implementation is that it is the first step in creating a manageable process. Once you have the initial process documented, you have a benchmark for measuring improvements. Ideas like designing equipment modules in parallel (for example, hydraulics designed in parallel to electronics) can be translated into product development cycle-time savings. In addition, this documentation for refining the development process can also be used as a project management tool to track ongoing design efforts. Finally, most small firms do a poor job of tracking the resources and costs associated with new product development. Characterizing the process enables you to also map out resource expenditures along the way. This understanding of the full cost of product development is valuable in determining profitability as well as in determining equitable billing rates for custom work.

Initiate the New Product Development Process as Soon as Practical

As shown by Figure 5.1, identifying a customer's need for a new product initiates the product development process. Traditional concurrent engineering tools designed to minimize/optimize the product/process development cycle are initiated once the concept for a new product is obtained. *Early identification of a new product concept using market orientation/customer orientation techniques enables an early start of the product development cycle, resulting in improved agility to deliver new products.*

Previous studies have found that market research is the most serious deficiency in the entire new product development process (Cooper 1986). The establishment of the Malcolm Baldrige National Quality Award is providing increased emphasis on market orientation/customer orientation issues with 300 of the 1000 points available in the evaluation criteria being placed on customer satisfaction (O'Neal 1992). Basic essentials of market orientation tools/techniques for early detection of new product needs include taking steps to make new product development efforts part of everyday business (Figure 5.2) and pursuing methods for getting new product ideas (Figure 5.3).

The five steps for making new product development efforts part of everyday business, identified in Figure 5.2, begin with understanding how new products relate to profitability (steps 1 and 2). Profitability can be enhanced by being able to offer entirely new products (step 1) or by modifying existing products to avoid losing your market niche (step 2).

Step 2 addresses the not uncommon situation of a smaller firm having its entire product line based on a single concept or technology that achieved the firm's initial success and established the market niche. Periodically, it is prudent for these firms to review technological advances to see if their product is in danger of being rendered obsolete. This is particularly true for electronics-based products. Keeping abreast of emerging competitors and associated product features can also spark product development efforts by seeing how desirable features can be incorporated into existing products. Finally, periodic reviews of dual-use opportunities can lead to new markets and products. For example, one firm discovered that an embedded equipment station in

---

1. Review your business plan to determine the importance of developing new products in regard to your company's profitability.

2. Review your existing products in terms of identifying changes needed to maintain market niche and/or avoid obsolescence.

3. Assign new product responsibility to an individual.

4. Provide guidance on budget and time to be devoted to new product development.

5. Implement a reporting system to track the status of new product development efforts.

Figure 5.2. Five steps for making product development part of everyday business.

1. Sales force
   a. Knowledge of customers' needs
   b. Inquiries from customers or prospects
   c. Knowledge of the industry or competition
2. Research and engineering
   a. Original or creative thinking
   b. Testing existing products and performance records
   c. Accidental discoveries
3. Other company sources
   a. Suggestions from employees
   b. Utilization of by-products or scrap
   c. Specific market surveys
4. Outside sources
   a. Inventors
   b. Stockholders
   c. Suppliers or vendors
   d. Agents
   e. Ad agencies
   f. Customer suggestions

Figure 5.3. Sources of new product ideas (Peter and Donnelly 1988).

its product could be broken out, modified, and sold as a stand-alone unit. This broadened the firm's product line, reducing the risk and dependency on its previous products.

Whereas many concurrent engineering product development techniques tell you *how* to develop new products, completion of these steps tells you *why* you should be developing them. Understanding this *why* is important for determining how much time and money should be devoted to the product development process. Step 3 assists in making sure new product development happens by making someone in charge. Step 4 provides guidance on how much that someone can spend, in terms of time and money. Finally, step 5 implements periodic status reporting on new product development efforts to keep things moving.

Sources of new product ideas, shown in Figure 5.3, are all steps that can lead to early detection of customer needs. These steps have the greatest potential for reducing the duration of the product development loop presented in Figure 5.1. Using these steps to detect a new product concept a few months earlier than usual will save significant time in getting a new product to the market. This time saved will be substantially greater than the time saved by applying advanced concurrent engineering techniques.

In addition to initiating new product development as soon as practical, a key aspect of this stage of the product development process is to make sure that the *right amount* of product change is pursued. Typically, the more significantly different the new product is from existing product lines, the more it will cost to develop and the longer it will take until shipment of the first unit. Careful cooperation between sales, engineering, and manufacturing (oftentimes only a few people in small firms) is essential to make new products a success. It is not uncommon to have the frontline sales force making promises to prospective customers regarding the attributes of a new product that translate into significant development time and cost requirements. Many of these development time and cost requirements can be avoided if the sales force is educated in terms of the time and cost impacts of design decisions. For example, one injection molding firm faced significant mold design and fabrication efforts to accommodate minor changes in a new product design. If sales personnel are aware of the cost and time implications of minor design decisions associated with a new product, they can focus more on what the customer needs to have from the new product versus ancillary features that represent significant costs to change.

Use Design for Manufacture Tools to Reduce the Design Effort

Once customer needs for a new product have been identified and an overall marketing assessment completed, product development efforts can be initiated, as shown by Figure 5.1. Design for assembly (DFA) tools to promote good design have been developed and primarily applied to the part and subassembly level (that is, micro level) of product design. The intent of these DFA efforts has been to make sure the design is easy and economical to manufacturing. These DFA tools and techniques are equally applicable to assist in improving the agility of the product development effort. This can be accomplished by using DFA concepts at the product level (that is, macro level) to reduce the amount of design work required.

The most widely used collection of DFA techniques has been packaged by G. Boothroyd and P. Dewhurst (1993) in the form of a handbook as well as software. This collection presents a structured approach evaluating the design to reduce the number of parts and to facilitate ease of assembly. A core set of DFA rules and how they might be applied to reduce the amount of design effort required at the product level is as follows.

• *Develop a modular design* (Suh, Bell, and Gossard 1978). A modular design provides two advantages. First, by designing a product with distinct components or modules where a component/module is designed to meet a specific functional requirement, you minimize the effect that a change in functional requirements or design solution to meet the requirements has on the total product design (that is, changes are isolated to a specific module). Secondly, the use of modules with standardized interfaces facilitates the use of standard components and existing process plans.

• *Design for a minimum number of parts.* Parts that are in fixed position to one another and can be made of the same material are prime candidates for combining into a single part. The benefits of combining parts (for example, fewer fasteners, ease of assembly) need to be compared to the added complexity of fabrication. Each part that can be reduced is one less part that has to designed and/or specified, shortening the design effort.

• *Standardize part types.* Standardization of parts, such as fasteners, structural components, and controls, reduces the number of stock items needing to be maintained by the manufacturer, reduces the need for customized components, and promotes larger buys from fewer vendors. This promotes designers being able to use off-the-shelf solutions, which will further reduce design efforts.

• *Design parts/modules for multiple uses.* Designing parts for multiple uses within a single product (for example, structural member serving as an alignment guide) and across several products (for example, housing used in multiple product lines) reduces the number of stock items and promotes more economies of scale in part manufacturing. The group technology use of part classification and variant planning (that is, modifying an existing design) are excellent methods for minimizing distinct part types and reducing the design effort.

The additional DFA technique of minimizing separate fasteners supports the manufacturing process but may result in additional product development time if not carefully managed. This DFA technique calls for combining fasteners and the use of snap fits. Separate fasteners require additional part management and added assembly operations. This DFA concept trades off more design complexity (possibility increasing design time) with greater ease of manufacturing. One plastic part manufacturer greatly simplified its product assembly operations using this technique, but at the cost of very complicated injection mold designs. The cost and development time trade-offs associated with part count reduction need to be understood to avoid moving problems from the manufacturing process to the design process.

At the product or macro level of designing, developing a modular design means isolating major assemblies from change. For example, equipment frames can be designed such that an increase or decrease in frame size can be accomplished by how much is bolted together, versus entirely new frame designs for different products. This concept supports standardized part types at the macro level because it enables one to use as much as possible of existing design, minimizing major assembly new design propagation. What this concept means is that any one particular product may have unused features (for example, bolt holes, mounting brackets). Although designers often have trouble with this concept, the net result can be multiple product lines that share the maximum number of major assemblies as practical. This approach can significantly reduce the time to develop "new" products, because new products are often nothing more than existing products with new features.

It is easy for designers to dismiss existing parts and subassemblies as being adequate for new products, as new features can usually be specified that render existing parts and subassemblies obsolete. To minimize part propagation, firms need to push for variant planning (using existing designs) of new parts and assemblies as much as possible. Short of computer-aided process planning packages (CAPP), this can be accomplished at the small-firm level by simply requiring a review of existing drawings as part of the formal design process. The more parts the company has, the more difficult it is to see if existing parts can be used. Sophisticated part classification systems, where the physical attributes of the part are captured in the part number, tend to be more complicated than what most firms need. A more effective strategy for managing part numbers is by having a sound bill of material structure that drives a similarly structured part drawing index. This enables designers to sort through existing bills of material to identify parts that could support the new design, and then easily retrieve associated part drawings.

Sound bill of material structures require that part information be separated from product structure information. Part information includes part number, description, cost, suppliers, and so on. Part numbers can be issued sequentially and have no meaning using this approach. Product structure information (that is, bill of material information) includes where-used or what-goes-into-what information, as well as when the part configuration takes effect. Part drawing indexes can be developed on top of the product structure by adding suffixes to indicate drawing types (wiring, drill and tap, assembly, and so on). PC-based bill of material packages are widely available, and most database products are readily adapted to support good bills of material applications.

Separating the file of part descriptions from what-goes-into-what product structure information avoids the trap of embedding your product structure into your part number, a habit that quickly leads to unmanageable systems. As an example of the problems associated with embedding product structure information into part numbers, consider a case in which the first three digits of the part number denote what product the part goes in and the last three digits represent the individual part. If the same part is used in a different product, two different six-digit part numbers represent the same part.

Designers may balk at using existing designs with little or no change when these existing designs could be modified to match "perfectly." However, the design process must at least enable the possibility to meet a new design requirement faster and cheaper with an existing design that is "close" versus a perfect design that will require significant additional time and money to develop. This phenomena of designers striving for the perfect design is particularly apparent in areas of design performance. It is not uncommon to have equipment designers striving for a few percentage points of performance improvement, while sacrificing the cost attractiveness of the equipment as a whole (as well as significantly increasing development time).

Designing parts/modules for multiple uses at the macro level attempts to enable designing in as much flexibility as possible. Opportunities to design in flexibility particularly exist in the area of onboard electronics and process control. Similar to modular design concepts, generic process control and onboard electronic capability can be adapted or tailored to meet specific customer needs. Thus, essentially new products can be produced that are nothing more than existing products that are programmed to behave differently. This strategy is not uncommon to the strategy used by PC manufacturers, where basic PC units can be used for individual workstations, network servers, and even process control equipment, depending on how they are configured. One problem that product designers seem to have with using commercially available software and/or process control products to achieve product "soft" flexibility is that often many features are not used. The tendency is to build from scratch just the right amount of software/process capability needed to meet design requirements. It is important to examine cost and time trade-offs between using general purpose software and process control products that are tailored for a specific product application versus developing/buying unique software and process applications that only meet the need of a single product. (See chapter 6 for additional discussion of technology to support flexibility in embedded software.)

Pursue CAD/CAE/CAM Tools/Techniques That Make Sense

New product development in larger firms makes extensive use of computer technology to complete the product development loop shown in Figure 5.1. Smaller firms often strive to make similar computer technology investments, with mixed results regarding enhancement to the firm's competitive posture. It is as if many of these smaller firms are trying to buy their way out of an inferiority complex. It is important for smaller firms to distinguish between investments that add to their agility in new product development versus investments that only add to the cost and complexity of their development process.

Intuitively, product development computer technology in support of a concurrent engineering process is appealing. It provides the foundation for information exchange throughout product/process development. This is particular true as CAD, computer-aided engineering (CAE), and CAM tools are employed. The common linkage between CAD/CAE/CAM applications is the solid model capturing volumetric information (Cassista 1992; also see chapter 6 for additional discussion of solid modeling). The solid model supports

- Realistic photo-quality CAD renderings that essentially represent a prototype of the finished part

- CAE programs modeling physical, thermal, and fluid dynamic processes

- CAM systems that translate solid model CAD data into code to directly drive CNC machine tools and other equipment

Additional elements supported by the concurrent engineering computer technology include the following:

1. Immediate access to information about previous product or process designs

2. Immediate access to information about manufacturability, reliability, maintainability, safety, performance, and other features related to elements of a design

3. Access to the most current state of the product or process configuration description as it is being developed within the multifunctional design team

4. A development tract outlining the rationale that went into developing the most recent version of the design

5. A notification capability that will flag design or manufacturing process changes

6. Data collection points to support the capture of design and implementation data

What computer technology investments make sense for smaller firms depends on each firm's individual situation. However, there are some rules of thumb that will help smaller firms avoid common pitfalls. For example, many of the benefits of large, integrated CAD/CAE/CAM systems designed solely for new product development can be achieved by common office automation that meets everyday business application needs. For example, PCs can support CAD applications that are adequate for most drawing needs of smaller firms. Small office networks are effective for routing design information and action items associated with new product development. The strategy of using common office automation resources to meet product development needs avoids the situation often encountered in larger firms, where specialized high-end workstations and software purchased solely for product development often have extremely low rates of utilization. The strategy of using existing and/or upgraded office automation resources also takes advantage of employees' understanding of existing computer hardware, resulting in reduced training costs.

Institutionalize Product Development Tools/Techniques

Once agile product development practices are identified and proven, they need to be institutionalized within the product development loop or system. Essential elements of an effective approach for institutionalizing product development best practices in smaller firms are as follows.

1. Establish a product focus team that includes

    —Marketing

    —Production

    —Engineering

—Support

—Purchasing

2. Design a concurrent process

3. Track execution

The formation of product focus teams in smaller organizations is not equivalent to forming committees where members often delay decision making (Pennell et al. 1989). Instead, the intent is to create a team designed for rapid identification and solution of product/process problems. This has to be carefully managed to ensure that team members' loyalty is primarily to the product and not to their functional departments (Wilson 1991). Also, creating a team does not mean that team participants reside in one place, although this is much more achievable in a smaller firm. A "virtual" team consisting of participants in separate locations can be created by

- Embedding concurrent engineering design rules in the form of checklists, databases, and knowledge-based systems to represent the interests of functional groups (for example, purchasing) to reduce the need for personnel communication between team participants

- Providing the infrastructure for effective product and process data information exchange between team participants

Both of these items can be supported readily by simple office automation in the smaller firm.

Tracking execution of the concurrent engineering process is essential to institutionalize the approach as well as providing the information base for process improvement. Key tracking parameters include cost and cycle time for the major product development steps shown in Figure 5.4, as well as some measure of the complexity of the product/process design effort. This information will improve time/cost estimates for future product/process design efforts and will identify areas that are higher-priority candidates for additional cost and time reductions. The use of the communication infrastructure that supports the cross-functional concurrent engineering team to track project execution

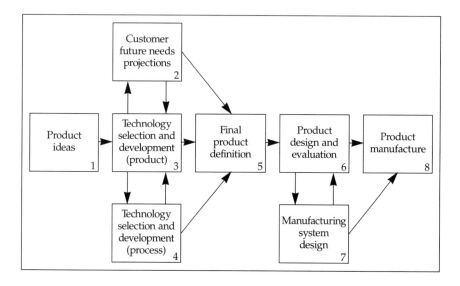

Figure 5.4. Concurrent process design (Wilson 1991). Reprinted from *Concurrent Engineering*. New York: Auerbach Publications. ©1991. *Warren, Gorham, & Lamont. Used with permission.

assists in team adherence to the process steps and serves as an ongoing indicator of project status to management.

Cycles of Learning

A final consideration regarding the pursuit of the agile product development loop is captured by the concept of cycles of learning. Each time a design, prototype, or product is developed, valuable insight is gained as to its effectiveness. The faster the development process, the more iterations or cycles of learning on the design can be completed. This ultimately will result in a better product than products manufactured by nonagile firms with more limited opportunity for design iteration. Having more opportunities to debug a new product before it hits the streets will do a lot to promote user acceptance and unnecessary warranty costs. An internally developed evaluation form to capture lessons learned from this process may be of value in future product development efforts.

Manufacturing Agile Practices

A number of large firms have been pursuing agile manufacturing practices using names such as world-class manufacturing and lean manufacturing. The goals of these efforts include maintaining market share, improving profitability, and improving the firm's ability to compete in a global marketplace. Smaller manufacturing firms have more down-to-earth reasons for pursuing agile practices when manufacturing product. For example, small manufacturers have few large production runs of standard product. This results in product changeover and retooling needs being proportionally higher than in larger firms. In addition, firms of fewer than 500 people simply do not have the cash flow to tie up large amounts of money in work-in-process inventory that results from nonagile work practices.

Agile practices can help smaller firms pursue high-mix, low-volume production while working within checkbook constraints. These practices consist of first establishing a manufacturing design, embedding the design into the shop floor layout, and implementing visual methods of controlling the manufacturing processes. Additional agile practices include striving for flow production and making sound make-versus-buy decisions. These practices are discussed in the following sections.

Establish a Manufacturing System Design

One of the most common mistakes that both large and small firms make is not having a well-understood manufacturing loop, as shown in Figure 5.1, that is communicated throughout the shop. A common nonagile manufacturing system consists of throwing production orders at the shop, unloading raw material into the front door, and hoping that a finished product comes out of the back. To add to a sense of internal control, firms place all sorts of tracking, reporting, and expediting procedures onto the shop floor to make the manufacturing system "work better." What is usually missing in these organizations is an established manufacturing system that lays the foundation for how things should work. Establishing a well-understood manufacturing system based on agile practices helps eliminate the need for the tracking, reporting, and expediting efforts that smaller firms cannot afford (larger firms cannot afford these costs either, but it is easier for them to absorb/bury these unnecessary costs).

A manufacturing system represents a business model that captures how the manufacturing system should work. In a perfect world, manufacturing systems would be simple and would consist of each order being translated into a shipped product with no wasted effort and no delay. In the real world, there are not enough people and material to work on orders as soon as they come in, and work must be prioritized, scheduled, and coordinated. Establishing a manufacturing system that reduces the dependence on non–value added tracking and expediting activities while designing for shop floor agility must be understood if the shop is to be effectively managed.

The first step is to design and document how the manufacturing process should work. This foundation for establishing an agile manufacturing system needs to capture when material is ordered, how work order priorities are determined, and how and when material is moved from workstations. An excellent method of documenting the manufacturing strategy is known as triple diagonal modeling (Shunk, Sullivan, and Cahill 1986; also discussed in chapter 3). While the particular method for mapping out the manufacturing strategy is not important, what is important is that all key elements of the manufacturing system be captured. There are three key elements that must be captured.

1. How production flows through the shop

2. How material and work processes are controlled

3. How production and resources are planned

The design of production flow through the shop needs to take into consideration the situations that personnel are going to encounter, such as what to do with material awaiting parts or material failing inspection. The product changeover process also needs to be designed. This enables personnel on the floor to know how to efficiently set up for new production runs and/or changing requirements. Rather than letting the shop floor personnel make these decisions haphazardly, a good manufacturing system identified in advance can assist in making sure the entire production flow process is executed in an efficient manner. The alternative is the "herding cattle" approach, where production orders, like cows, meander around on the shop floor until expediters herd them out of the door.

Understanding and documenting how the material and work processes are controlled means that one identifies what should be going on, collects information on what is going on, and takes corrective actions. Typical control elements are

- *Schedule*—Comparing what is scheduled to be produced versus what is actually being produced and reallocating resources as required

- *Quality*—Comparing quality requirements to actual quality and adjusting the manufacturing process as required

The main advantage in designing the control strategy is that it identifies the essential information needing to be monitored. This information is often significantly less than the large amount of information collected in many production environments. By tying the data collection into the control strategy, many firms can answer the questions "Why are we collecting this information and what are we going to do with it?" for the first time.

Once the manufacturing system design has been established, the foundation for agile manufacturing has been laid. First, non–value added activities have been largely eliminated, reducing the amount of delay and expense associated with many manufacturing enterprises; second, a formal manufacturing system has been identified that determines how changing production requirements are to be met. The next step is to execute the manufacturing system, which is described in the following section.

Embed the Manufacturing System Design in the Shop Floor

A winning strategy for making sure that the agile manufacturing system is executed is embedding the manufacturing system design in the shop floor as much as possible. What this means is that production area layouts and material handling equipment are marked, configured, and positioned to communicate the desired manufacturing system flow to shop floor personnel. Pallet set-down areas can be marked on the floor to manage input and output queues. Special set-down areas can be created to handle quality problems and for assemblies waiting for parts. The key is to create a shop floor environment that enables workers to automatically know where material should go next. This enables workers to execute the manufacturing system as designed, with a minimum of unnecessary tracking and reporting systems. For example, one firm outlined its material flow lines by configuring overhead lights to match how material should move across the shop.

Agility can be incorporated into the shop by making both the layout and the material handling equipment flexible. For example, one firm achieved agility by replacing roller conveyors with rail-guided carts. As the production mix between products changed, production lines could be created or removed by moving the rails on the floor. If more workstations are to be added to the production line, it can be lengthened by adding more carts and railing. Work-in-process can be controlled by car availability. All incoming and outgoing queues can be marked. The use of carts as the material handling method of choice has many advantages. Empty carts make natural kanbans or signals that more production is needed. Carts also can be designed to provide a kiting function. Special compartments and hooks can be added to the cart and labeled to assist in making sure all required material in support of assembly is collected.

Implement Visual Methods of Control

Similar to embedding the material flow strategy of the manufacturing system in the shop layout and material handling equipment, visual methods can be used to perform control strategy (Greif 1991). Effective visual control strategies that have been observed include

- Using large white boards adjacent to work cell areas to track scheduled versus actual production
- Managing part mix by using the number of pallets in designated pallet set-down areas to prioritize what gets worked on next
- Having green, red, and yellow lights indicate if a work cell is busy, idle, or down due to maintenance
- Having large digital displays showing workstation production volumes mounted high on the walls, used to assist in line balancing

- Having defective versus good parts mounted on boards to assist in inspections
- Establishing two-bin inventory systems representing physical reorder points
- Using colored shop tags to denote job order due dates
- Using color-coded work cells and corresponding shop tags to assist in routing
- Having status boards that capture the time work entered a manufacturing cell, expected time out, and manufacturing problems, if any

The advantage to pursuing a visual control strategy is that it is significantly cheaper and easier to understand than computerized shop floor control systems. In terms of agility, visual control strategies are also often much easier to modify than computer-based systems.

Strive for Flow Production

The concept of flow production means that individual production items flow though the manufacturing process individually, without batching and unnecessary routing. The capability to support one-piece flow eliminates the delays associated with batching. These delays are caused by production units in batches or lots sitting idle in front of a manufacturing process while part of the batch or lot is being worked on. One-piece flow, if properly executed, can also eliminate delays associated with routing. Routing delays are common in job shop environments where production units have to be extensively moved from one manufacturing workstation to the next. Reducing batching and routing delays results in a short manufacturing lead time. Advantages of a short manufacturing lead time are numerous and include the following.

- *Reduces work-in-process dollars.* The amount of dollars tied up in work-in-process (WIP) is equal to:

$$\text{Manufacturing cycle time (days)} \times \text{daily production} \times \text{average value of WIP}$$

A common first-order approximation for the average value of a production unit in WIP is half the cost of a finished piece of production. This estimate is usually low for most situations, as material cost is often the largest expense and is usually purchased at the beginning of the production cycle. Better estimates as to the average value of a unit of production in WIP can be obtained by summing the entire material cost with half the labor cost. Regardless of the specific costs, the expression displayed earlier shows that every percent reduction in manufacturing cycle time results in a percent reduction in WIP costs. Moving to one-piece flow by eliminating batching and other unnecessary delays can significantly reduce work-in-process dollars.

• *Supports building to order versus building to stock.* The longer the manufacturing cycle time, the longer a company has to see into the future to determine production requirements. If the manufacturing cycle time exceeds the lead-time expectations of customers once an order is placed, firms have to forecast (that is, guess) production orders and produce units to stock before orders are in hand. This creates a finished inventory system and associated costs, and puts firms at risk in terms of over- or underproduction. Reducing manufacturing cycle time to less than or equal to customer lead-time expectations supports building to order, significantly simplifying the production master scheduling process and eliminating finished inventory costs.

• *Simplifies shop floor control.* The less material on the shop floor to manage, the easier shop floor control becomes. As reduced manufacturing cycle times eliminate WIP, fewer production units will be in the shop. This reduces tracking transactions and shop floor inventory management activities. Additional benefits include floor space requirement reductions resulting from fewer WIP production units doing more moving and less sitting.

As manufacturing firms increase in size, they evolve in regards to how material flows through the shop. Initially, a common material flow strategy for smaller firms is bay-style production. Bay-style production is where all manufacturing steps are completed in the same area (very similar to making something in your garage). More and more parts are routed around the shop as more equipment in support of the manufacturing process is obtained and more parts are manufactured in-house.

This evolves into a job shop environment with extensive part routing supported by bay-style assembly areas. Finally, as production levels of more standard product increase, the equipment is integrated into cells and lines, creating a flow shop. Advantages of a flow shop include

1. A reduction in part routing and associated delays

2. Greater ease in managing production (one can see things flow)

3. Better ability to get manufacturing workstations to standardize on best practices (lines can become specialized)

4. Easier to keep individual manufacturing processes in balance in terms of production rates

5. Easier to migrate to one-piece flow because delays are more apparent and can more easily be targeted for elimination

The challenge for small firms is to accelerate the job shop to flow shop evolution to gain flow shop advantages as soon as possible. To accomplish this, small shops need to devise a strategy to adapt their low-volume/high-mix production environment to resemble the more standard product/higher-volume production mix of larger shops supporting smooth flow of product through the manufacturing cycle. To do this, the diversity in parts needs to be dealt with by grouping similar parts or products together and by minimizing the effort to go from working on one part/product type to the next. This can be accomplished by using the following agile techniques.

Agile Technique: Group Technology Methods for Small Manufacturers
The first step in grouping similar parts or products together to support moving from a job shop to a flow shop is to identify equipment and workstations required by each product (Sekine 1991). This approach is called *production flow analysis* and is a graphical technique for completing a group technology analysis. To make the analysis more manageable, it is important to focus on the products making up the majority of production. This information can be obtained by listing the last year's production totals and ranking them in order of quantity produced. An example analysis is shown in Table 5.1. The group technology analysis should initially be focused on the few products representing most of the

Table 5.1. Rank order analysis of production.

Product	Percent of quantity	Cumulative total production	Percent
Model A	475	35.9	35.9
Model B	312	23.6	59.5
Model C	175	13.2	72.7
Model D	135	10.2	82.9
Other	225	17.1	100.0

production. The analysis can be expanded once initial flow lines have been established.

Routing information needs to be obtained for each of the initial products to be transformed into flow production. A production flow analysis is constructed by listing the manufacturing workstations and equipment identified by the routing along the top axis (columns) and individual products listed down the vertical axis (rows). Whenever a product routing requires a workstation or a specific piece of equipment, the corresponding column is marked. An example production flow analysis chart is shown in Figure 5.5.

Production flow analysis charts enable one to complete a group technology of products by inspection if the number of products, pieces of equipment, and workstations are not too high. More complex situations require the application of mathematical clustering techniques. It should be mentioned that there are other approaches to group technology based on part classification systems (Groover and Zimmers 1984). However, part classification–based group technology approaches usually exceed the needs of smaller firms.

A visual inspection of Figure 5.5 (a very simple example) reveals that all products must flow through workstation 1. Next, product models A, B, and D have similar workstation and machine requirements and can probably be grouped together (product group 1). Product model C could be treated separately (product group 2), resulting in two major product flows following workstation 1. As both flows

Products	Work-station 1	Work-station 2	Work-station 3	Machine X	Machine Y	Machine Z	Product Group
				Shop Resources			
Model A	●	●	●		●		1
Model B	●		●				1
Model C	●			●	●	●	2
Model D	●		●		●		1

Figure 5.5. Production flow analysis approach to group technology.

require machine Y, two are needed (one for each flow); otherwise, the two product flows need to be near enough to each other to share machine Y.

The production flow analysis is useful for identifying workstation and equipment to support product groupings, but it does nothing to look at the routing of a product to determine how workstations and equipment should be arranged. This issue is addressed by the concept of manufacturing cells.

Agile Technique: Manufacturing Cells

Once product groupings and associated workstation and equipment requirements are established, routing information can be used to lay out the shop to support smooth product flow. The concept of manufacturing cells solves many of the layout problems associated with trying to accommodate different part/product routings within a product group, because workstations and equipment can be organized in a U-shaped configuration or a cluster (as shown by Figure 5.6). This makes it easier for workers inside the U-shaped cell or cluster to access any particular workstation or piece of equipment. Therefore, if a particular routing calls for a piece of equipment to be skipped or the part has to backtrack, it is easier for the worker to accommodate the flow of material than if the equipment was configured in a straight production line.

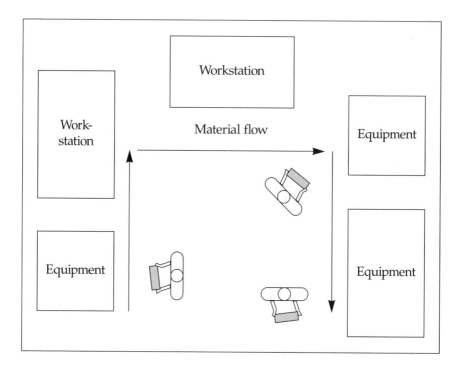

Figure 5.6. U-shaped manufacturing cells.

Cell size should be limited to some number of workstations and equipment, configured in a cluster or a U-shape, where workers are within several steps from any one workstation or piece of equipment to the next. This enables workers to move easily from one area of the cell to another to accommodate different part routings or to enable one worker to operate multiple workstations and/or pieces of equipment. A first approximation as to the manufacturing cell layout can be obtained by working with a part or product representing a higher–production volume item. Ideally, this higher-volume item also uses most of the workstations and pieces of equipment associated with the product grouping to which it is assigned. The routing and processing times for this item need to be obtained and each work step listed in order as identified by the routing, as shown in Table 5.2.

Table 5.2. Using routing information for manufacturing cell design.

Routing step	Process name	Time (minutes)
1	Workstation 2	20
2	Workstation 3	15
•	•	•
•	•	•
•	•	•

Forming a first cut at a manufacturing cell consists of going down the routing list until you have a reasonable number of workstations and equipment to form a cell that is of acceptable size. Workstations and equipment can then be placed in a U-shape, as shown by Figure 5.6, in order of the routing for the part or product under consideration. This provides for a smooth flow of material through the manufacturing cell. Depending on the number of workstations and pieces of equipment in the routing, multiple manufacturing cells may be needed to keep the cells a manageable size.

After the first cut of the manufacturing cells is obtained, the routings of the other items in the same part or product family need to be considered. These routings may contain additional workstations and pieces of equipment not included in the first cut design. Also, the routing sequence between workstations and equipment may be different. The additional work stations and equipment should be inserted into the rough cell design to minimize backtracking in the flow. Completing this exercise by working with paper templates is highly recommended. In cases where backtracking is required, the cell layout can be changed to see if a net reduction in backtracking can be obtained by changing the sequence of workstations and equipment.

Once the cell design is modified to accommodate all of the items in the part or product group, cell balancing needs to occur. Cell balancing refers to designing the manufacturing cells and cell staffing levels such that all cells contributing to a single product group have about the same rate of production. This is similar to line balancing in traditional assembly lines. Cell balancing is required because the cells will operate in parallel, and out-of-balance conditions will result in queues in front of some cells and/or idle cells. Cell balancing can be achieved by calculating the daily throughput rate of cells and adjusting cell content and/or staffing until all cells have similar rates. For example, removing the last workstation or piece of equipment from a cell with a low throughput rate to the beginning of the next downstream cell is one balancing technique.

Once a refined design for the manufacturing cells is available, a final revision to the cell design and determination of cell placement on the shop floor needs to be made by looking at the relative difficulty of relayout. Large pieces of equipment, heat-treating equipment, painting booths, cleaning vats, and other difficult to relocate items may drive a revision to the manufacturing cell design. These items, also known as *monuments,* may be uneconomical to move, requiring that the cell design be revised. (See chapter 6 for examples of technology that can overcome this problem.)

Once implemented, the manufacturing cells will better enable parts and products to flow through the shop individually (that is, single-piece flow), instead of being batched as in a job shop environment. However, there are some additional steps that need to be completed to further enhance the agility of the shop floor. These additional steps include mix model production and setup and changeover reduction.

Agile Technique: Setup and Changeover Reduction

A large number of techniques assist in reducing the time to setup for a product or to changeover between products (Shingo 1985). The techniques can be grouped into two major classes. The first class of setup reduction techniques divides setup into activities that can be completed while machines are operating (referred to as external setup), versus activities that require machines to be shut down (referred to as internal setup). Once the external and internal portions of setup (or changeover) are understood, external setup activities (such as getting the next fixture to be used) are completed during equipment operation, resulting in reduced production time loss due to changing over between parts or products. The next class of setup reduction techniques focuses on reducing internal setup times. This can be accomplished by using quick position hold-down and release mechanisms.

Know What Material Management Strategy to Use

Another key part of a firm's manufacturing system is to select the material management strategy that supports the firm's agility requirements. Given the nature of most smaller firms' product lines, many firms do not build product to stock, but rather build product to fill specific orders. *Building to order* means that no work is done or material procured until an order is in hand. Building to order is usually the preferred approach for smaller firms if the manufacturing lead time is small enough to accommodate customer requirements. The reason it is the preferred approach is that it ties up less cash in inventory. If faster manufacturing lead time is required, an *assembly to order* strategy can be pursued. Assembly to order means that all basic components are fabricated or purchased prior to receiving an order, and when an order is received, these components are combined into a unique combination for that customer. The purpose of bringing up building to order versus assembly to order is that it drives the whole inventory control strategy. For example, when building to order, inventory is ordered and controlled for a specific number of end items. With assembly to order, however, one has to manage inventory levels and allocations to specific orders from these levels. There are advantages and disadvantages to both strategies.

Have a Make-Versus-Buy Strategy That Makes Sense

Make-versus-buy decisions (that is, deciding what is produced by the manufacturing loop shown in Figure 5.1) for component parts also have the potential to be improved in many smaller manufacturing firms. One ineffective strategy often encountered is to buy as many components as possible and focus solely on assembly and packaging. While this minimizes capital and skill requirements, it leaves the firm at the mercy of vendors in terms of part price and availability. Firms often overlook the benefits of having more in-house fabrication capability. In-house fabrication capability usually supports agility, because new designs often can be translated into parts faster than if the design is forwarded to vendors. Also, fabricating parts in-house can provide valuable feedback regarding the manufacturability of the design. This feedback is more difficult to obtain from vendors, as vendors are sometimes reluctant to identify cost-cutting opportunities. In-house fabrication capability also tends to support research and development, as well as fabricating prototype of new parts and products, which also can be valuable. Finally, the ability to make parts in-house also contributes to workforce load leveling. During lulls in the production schedule, more parts can be made in-house.

Another ineffective make-versus-buy strategy is to fabricate as many components in-house as possible. While this may give the firm maximum control of parts, it can limit agility because the firm's ability to produce some parts in-house may be inferior to vendor production capability. One firm using the make-everything strategy took longer to fabricate in-house process control equipment that had fewer features than readily obtained and adaptable vendor products. It was just too difficult for the firm to compete effectively in the area of process control design and fabrication, which represented a small part of its production.

Final Considerations

Additional considerations that can assist in moving firms toward agile manufacturing practices include the following.

Understanding the value of investing in people. Larger manufacturing firms can more readily support the concept of division of labor (that is, workers completing a narrow set of duties) than smaller firms with much smaller headcounts. This creates a situation where employees need to be more versatile in smaller firms than in larger firms. Coupling this with the fact that salaries and job security are often lower in smaller firms results in a situation where smaller firms need people to do more work for less money. Understanding the value of investing in people in terms of resources contributing to a firm's agility is very important. Knowledgeable, cross-trained employees improve the responsiveness of the shop floor at a much more cost-effective rate than machining centers. Understanding that workers have a much greater impact on cycle time than most process equipment is an important first step in recognizing the need to invest in personnel.

Removing inappropriate fear from the shop floor. One phenomenon encountered in smaller manufacturing firms that is encountered less often in larger firms is micromanagement of functional areas by the top manager (usually an owner). These managers have not reached a stage where they can admit they do not have time to do everything and can relinquish their control to subordinates. Given that firings often can happen more readily in smaller firms than in larger firms with more established termination procedures, individuals may be less likely to point out better ways of doing business that conflict with the direction of top management. Progressive firms pursuing agility need to eliminate this phenomenon. (See chapter 10 for more on management.)

Not sacrificing agility for perfectionism. One final thought regarding implementing change that can lead to agility is the concept of not waiting until the perfect solution exists before making change. Because change is an evolutionary process, enhancements can always be incorporated within the next change iteration. The lessons acquired through rapid partial or full implementation of agile concepts often cannot be identified prior to attempting to implement an agile practice. As George S. Patton said, "A good plan violently executed NOW is better than a perfect plan next week" (Patton 1947).

References

Boothroyd, G., and P. Dewhurst. 1993. *Design for assembly—A designer's handbook.* Amherst, Mass.: Department of Mechanical Engineering, University of Massachusetts.

Cassista, A. L. 1992. *Concurrent engineering. Best manufacturing practices.* LaJolla, Calif.: Office of the Assistant Secretary of the Navy.

Cooper, R. G. 1986. *Winning at new products.* Reading, Mass.: Addison-Wesley.

Greif, M. 1991. *The visual factory.* Cambridge, Mass.: Productivity Press.

Groover, M. P., and E. W. Zimmers Jr. 1984. *CAD/CAM computer-aided design and manufacturing.* Englewood Cliffs, N.J.: Prentice Hall.

Maddux, K. C., and S. C. Jain. 1986. *CAE for the manufacturing engineer: The role of process simulation in concurrent engineering.* New York: American Society of Mechanical Engineers.

McHose, A. 1992. *Manufacturing development applications.* Homewood, Ill.: Business One Irwin.

O'Neal, C. R. 1992. Good enough is no longer good enough. *Target* (Association for Manufacturing Excellence) (May/June): 15–20.

Patton, G. S., Jr. 1947. *War as I knew it.* Boston, Mass.: Bantam Books.

Pennell, J. P., R. I. Winner, H.E. Bertrand, and M. M. G. Slusarczuk. 1989. Concurrent engineering: An overview for Autotestcon. Presented at Autotestcon 1989: The System Readiness Technology Conference, IEEE.

Peter, J. P., and J. H. Donnelly Jr. 1988. Preface to *Marketing management.* 4th ed. Homewood, Ill.: BPI/Irwin.

Sekine, Kenichi. 1991. *One-piece flow: Cell design for transforming the production process.* Cambridge, Mass.: Productivity Press.

Shingo, S. 1985. *A revolution in manufacturing: The SMED system.* Cambridge, Mass.: Productivity Press.

Shunk, D., B. Sullivan, and J. Cahill. 1986. Making the most of IDEF modeling—The triple-diagonal concept. *CIM Review* (fall): 12–17.

Suh, N. P., A. C. Bell, and D. C. Gossard. 1978. On an axiomatic approach to manufacturing systems. *Journal of Engineering for Industry, Transactions of ASME.* 100 (2): 127–30.

Wilson, C. C. 1991. Potential pitfalls of concurrent engineering. *Concurrent Engineering* 1 (1): 37–43.

CHAPTER 6

Implementing Technology
to Enhance Agility

Lawrence O. Levine and Brian K. Paul

This chapter discusses some examples of technology that can enhance agility in manufacturing production systems. It emphasizes the necessary conditions that should exist prior to a major investment in new technology and explains some key guidelines for effectively implementing technology. It explores issues to consider when evaluating a technology make-or-buy decision. Finally, it provides suggestions for how small and midsized manufacturers can access assistance to take advantage of new technology.

The right mix of technology can make the organization more flexible in the design and manufacture of products. There are five fundamental reasons to adopt technology to enhance agility. Technology

- Reduces the product development time to market
- Reduces the product delivery time to the customer
- Improves workforce capabilities and flexibility
- Enhances the flexibility of the production facilities
- Improves understanding and control of production processes

Several other uses will not be covered in this chapter. For example, technology can be used to reduce or eliminate hazardous conditions for employees. It also can be used to eliminate tedious jobs and free up employees for more value-added activities (for example, hard automation and robotics). Because of the focus of this book, these important technology applications will not be discussed.

The question is not Should a manufacturer adopt high technology in the pursuit of agility? The real issues are how to find or develop appropriate technology and how to quickly and inexpensively deploy this technology. Agile technology is appropriate when it fits within a well-understood operational concept that consistently supports flexibility, level production flow, and short cycle times. However, technology should be adopted only where it demonstrates a clear connection to business objectives. For instance, if the business involves the mass manufacture of standard mechanical parts for several large firms, the business objectives would probably not require agile manufacturing; therefore, an acquisition of rapid prototyping equipment would not be in order. Similarly, if the business is a job shop to many local manufacturers, the acquisition of a manufacturing resource planning system with standard bills of material would not make sense.

Agile manufacturing technology should be adopted only when an organization understands how the technology will be used within its business processes to enhance agility. It should not be purchased simply as a stimulus for becoming agile. Otherwise, the organization will run the risk of causing major work disruption within existing production systems.

How to Implement New Technology—Beating the Odds

Before a manufacturer starts any major technology implementation effort, it is important to note that studies have found that 50 percent to 75 percent of U.S. companies failed in their attempts to implement advanced manufacturing technology (Chew, Leonard-Barton, and Bohn 1991). There are many reasons for this sorry track record, but underlying it is a common theme. Managers typically think of technology adoption as just another investment decision followed by a project to be managed using standard project management techniques (planning, budgeting, cost/schedule performance, and so on). Because agility is achieved by the linking of many elements, traditional investment analysis techniques may not capture the true value of a specific investment. Moreover, implementing significant changes in a firm's technology usually involves overcoming uncertainties in the capabilities in the new technology; integrating it successfully with the existing technology; and modifying the organization's business processes, structures, and incentives.

Traditional project management approaches can seriously underestimate the amount of research and iterative learning that must be managed to minimize the near-term negative consequences of technology adoption. Therefore, managing technology adoption should be a part of the larger effort to manage learning by the agile manufacturing organization—speeding up the process of learning while minimizing the cost and disruption that learning can cause. This goal can be achieved by following these guidelines, which are explained in subsequent sections.

- Manage technology adoption as organizational learning.
- Manage technology adoption as part of work process redesign efforts.
- Learn early, learn cheaply.
- Address organizational issues early.
- Provide education and training.
- Document and share lessons learned.
- Budget for negative impacts.
- Provide adequate resources given the scope of the change.

Manage Technology Adoption as Organizational Learning

Managers should make organizational learning an explicit objective of each technology project. Clear objectives should be stated concerning the improved performance to be achieved and the anticipated scope of the organization that will be involved in learning about the new technology. Estimates of the likely organizational impacts should be made before the start of each project. These estimates can be used to begin selecting appropriate people to be involved in the implementation team. Because of the uncertainties and difficulties that must be overcome during implementation, management should insist on frequent communication on project status. Project managers should not be expected to independently solve all problems discovered during implementation, but should uncover problems as early as possible to reduce the risk to the organization. More specifically, a mechanism should be established to quickly surface and resolve issues uncovered throughout the project. These issues will typically concern organizational impacts (for example, changes in roles/responsibilities, inconsistent performance measurement) and changes required during the implementation and operation of the new technology (for example, modifications to the new or existing technology). This mechanism should allow for broad functional participation so that clear broad-based decisions are reached before significant investments are made.

It is often appropriate to appoint an expert user, rather than a technical specialist, to the project management role. This user will have familiarity with the work environment. While this individual probably will need education on the technical aspects, this user will have deeper organizational and procedural knowledge. This insight will help the user be sensitive to potential issues. He or she can act as a scout for management to provide early warning of problems and a more realistic assessment of resource requirements (particularly training needs). The expert user is also likely to have more credibility than a technical specialist in "selling" the technology to coworkers.

Manage Technology Adoption as Part of Work Process Redesign Efforts

A major issue in the transition to agile manufacturing will be in maintaining integration among several program activities occurring concurrently. Work process redesign teams (discussed in chapter 4) will quickly become frustrated if their efforts conflict with technology implementation teams. Rather than isolating technology implementation teams from work process redesign teams, it should be recognized that a major output from work redesign efforts should be technology implementation. Technology implementation issues such as education and training, job redesign, and organizational resistance are inherently easier to manage if they are integrated with work process redesign.

The coordination of activities under one integrated program can be simplified by making one person responsible for coordinating all work process redesign and technology adoption activities. This program manager should work with upper management and project leaders to ensure that strategic direction and project feedback are adequately communicated. Program management responsibilities should include regularly reviewing the status of project activities, coordinating project team activities when necessary, requesting program budgets from upper management, and evaluating project leader performance.

Learn Early, Learn Cheaply

A key strategy for successful technology adoption is to increase the scope of learning while reducing the cost. This can be done in a number of ways. The first approach is to learn from the successes and problems of others. While no two organizations are exactly alike, much can be learned from the experience of others, particularly in avoiding difficulties. Potential lessons learned can exist within an organization or outside it. Both sources of learning should be used, as this can be the single cheapest way to learn. While consultants can be a source of useful information in this area, the lessons are often more likely to be taken to heart through person-to-person communication between technology-adopting organizations.

Often technologies will impact the dynamic processes (for example, shop floor operations) of a production system. Dynamic processes are very difficult to visualize for most people. Simple qualitative simulations using people to demonstrate operations can help highlight likely impacts of new technology and also help educate employees about the nature of the change. In addition, these human simulations (see chapter 3 for further details) can sometimes identify key activities that will require particular attention—activities that can significantly impact overall production system performance. These activities can then be targeted for more detailed research and development and/or staff education and training.

New technology must be installed in a particular physical work environment. While sophisticated CAD/CAM and ergonomic simulations can be used, often crude mock-ups and layouts can be effective in identifying problems and communicating with employees during the design process. Chalking lines in a parking lot and having employees "walk through" the proposed layout can help identify problems early. Employees will often accept crude representations of proposed designs when they believe their input is being incorporated in the design process.

For more exact analysis, computer modeling can help anticipate problems and predict system performance. Typically, staff experts are needed to build the model and interpret the results (generally consuming work weeks to months of effort). However, this can still be cost-effective in many situations where the impact of the new technology is likely to be significant. For smaller organizations without internal staff resources, nearby universities can often provide simulation services from professors and graduate students. For further information on the effective application of two types of modeling, discrete-event simulation and queuing, see Harrel and Tumay (1995) and de Treville (1992), respectively.

Many types of new technology involve interacting with computer systems. Rapid software prototyping techniques can quickly clarify user requirements when used effectively. Software prototyping can quickly create systems that look like and interact with people in ways that are similar to the systems firms need to build. The prototype can be an effective way to communicate with users about their needs and preferences, as well as give them an early feel for the new technology. The prototype also can serve to document the requirements to system developers, thus becoming a vehicle to share the learning that has occurred during requirements determination.

Address Organizational Issues Early

When technology projects fail, a frequent cause is failure to address the changes to the organization that are required to make the technology work. One way to help ensure organizational issues will be addressed is to check whether top management has a shared understanding of how the technology will support the organization's operational strategy. If this consensus exists, there is less likelihood that narrow functional views will obstruct technology adoption.

Second, those organizations most impacted should be represented in the design and implementation efforts. Functional representatives should be included on the project team to provide meaningful input.

Efforts should be made to educate these representatives on the technology to help broaden their perspectives. Many times technology implementation will be a natural follow-on to work process redesign efforts. As such, a well-trained, cross-functional team should already exist for technology implementation.

Third, organizational issues should not be ignored during pilot adoptions of the technology. A prototype organizational structure can be established and tested in combination with the technology to adequately evaluate the impacts to the entire organization. Evaluation of these softer issues should be made as explicit as the evaluation of the technical aspects of the prototype performance.

Provide Education and Training

Education and training are often afterthoughts when adopting new technology. The scope and depth of this requirement are often underestimated. Employees must be educated on why the organization wants to use the technology. They need to understand how this technology will fit into the significant processes of the organization. If the technology embodies any theories, methodologies, or algorithms, then these need to be taught and understood. When the operation of a technology is a "black box," employees will have limited ability to debug it in the near-term or improve it over the long-term. Much of this training and education should be addressed naturally within the context of work process redesign teams.

Different people learn in different ways. The education and training should allow individuals to use a variety of mechanisms to understand and become effective with the new technology. Reading, lecture, discussion, demonstration, and hands-on use should all be elements of effective programs. Broad education, concerning underlying theories or why the technology is being adopted, can start early. On the other hand, task-specific training, concerning how to use a particular technology product, should be provided as close to actual use as possible. Again, these issues should be addressed within work process redesign efforts. Consultants, technical societies, and local community colleges are all potential sources of education (books, videos, short courses) for both management and staff.

Document and Share Lessons Learned

Sharing lessons learned should be an explicit deliverable of each technology adoption project. Moreover, the organization should try to encourage the sharing of "war stories" as an explicit part of the culture. It is critical that the organizational climate allow failures to be shared as widely as successes. While successes may help achieve organization goals, avoiding failures by learning from the difficulties others have faced will probably have a greater cumulative impact on achieving and maintaining agility. This responsibility can be fulfilled by ensuring future work process redesign teams are staffed with veterans of past redesign efforts. In addition, management should task a central work process redesign program organization to facilitate communication among peers across the organization (particularly at dispersed locations) to speed the learning process and reduce the perceived risk in reporting problems.

Minimize the Negative Impacts on Existing Customers

The physicians' creed includes the injunction to "first do no harm." This may be too high a standard for many organizations to achieve. However, managers should devote adequate resources to minimize the impact of technology adoption on existing customers. This can be particularly important if some aspects of the new technology can only be understood when operating under real-world conditions. This can be accomplished in several ways. In some instances, old and new technology can be run in parallel until performance is verified. This option may be prohibitively expensive or physically impossible. Some technologies can be proven during off-shifts if the facility does not have continuous production. If the new technology must be run during normal operation, it should be explicitly scheduled and production targets should be reduced accordingly or overtime authorized to meet customer obligations.

Provide Adequate Resources Given the Scope of the Change

Adopting these guidelines implies the consumption of organizational resources. There are easier and harder ways to implement a new technology, but there are no free lunches. Sometimes an organization will be unable to expend the necessary resources to pursue these approaches. If this is the case, then management should examine if the technology can be implemented in phases to reduce the scope of the impact. Perhaps only some capabilities should be implemented immediately or the technology should be used on a portion of the facility's workload. A conscious management decision should be made to focus initial efforts where significant benefit can be gained while minimizing disruption. This strategy trades off longer time to implement for reduced resource consumption per period.

Preparing for Technology Implementation—A Checklist

While the strategies and techniques just discussed can minimize the pain in adopting technology, they are not sufficient to ensure the technology you implement will increase your organization's agility and flexibility. New technology must be built on a solid foundation to reduce the risk of implementation failure and disappointing results. Think of this foundation as a preflight checklist for review prior to making a major commitment to adopting new technology. The elements on this checklist include

- Is the operational strategy linked to the business strategy?
- Is the program integrated?
- Are current methods documented and standardized?
- Has the plant layout been improved with low-cost changes?
- Have machine setup times been reduced?
- Have key bottlenecks been improved?
- Have quality and preventive maintenance programs been established?
- Has mixed-model scheduling been established?
- Are non–value added activities being eliminated?
- Have simple automation opportunities been exploited?

Most of these elements should be underway or completed before major changes to technology are implemented. Each of these elements is highlighted in the following sections. For more details on these subjects see Harmon and Peterson (1990), Hall (1983), Schonberger (1986), and Shingo (1985).

Is the Operational Strategy Linked to the Business Strategy?

Doing the right thing must precede doing things right. For any organization to be successful, its business strategy must be directed to meeting real customer needs. Top management has identified and articulated this strategy to the rest of the organization and ensured there is real understanding and commitment to achieving it.

Next, management has a vision of an operational strategy that is consistent with achieving the business strategy. Management understands, at least in broad terms, how the operation of the production system will result in meeting customer needs and business goals (see chapter 3). This vision includes understanding how the planning, control, and execution of the work will link together in a system that can be managed without constant fire fighting and expediting. Finally, this operating vision has been shared and understood by everyone in the organization so that everyone knows how their work fits into the larger whole.

Is the Program Integrated?

A program manager has been made responsible for coordinating all transition activities and has been given the resources necessary to carry out such coordination. Work process redesign efforts are well underway. Redesign efforts are initially organized based on the strategic direction from management. New redesign teams are formed based on the recommendations of existing redesign teams. Project leaders are given responsibility for project activities on each team. Apprentices are chosen from among team members to support each project leader. The experience gained from prior efforts serves as leadership training for apprentices who are used to lead future redesign teams.

Redesign teams are used as the transition to continuous improvement within the various work environments. Job metrics and incentives are adapted to encourage continuous improvement as part of the redesign process. Technology adoption evolves as a result of work process redesign efforts. Technology project teams do not exist separate from work process redesign teams.

Are Current Methods Documented and Standardized?

Existing design and manufacturing work processes are understood by managers and staff alike. Process flows have been documented and this documentation is used in the training of employees. Standard procedures are documented and followed consistently. Workers are actively involved in the development and improvement of these procedures.

Have Bottlenecks Been Identified and Improved?

Key bottleneck operations in major business processes are identified and known by all staff that interact with them or impact their operation. Steps have been taken to ensure no bad inputs are received by bottleneck operations. In addition, efforts have been made to locate and quickly activate overflow capacity for these operations. Transfer lot sizes from bottleneck operations have been minimized if production lot sizes are still too large (for example, if lots must be made in batches of 1000, as each 100 are produced they are transferred to the next operation).

Has the Plant Layout Been Improved with Low-Cost Changes?

Products with similar process requirements have been identified. The layout has been modified to reflect a product orientation by establishing manufacturing and assembly cells that are placed in proximity to each other to reduce the amount of material handling. While major facility modifications have not been made, easy-to-achieve relayout has been accomplished. Unnecessary (non–load bearing) walls have been removed. Fixed shelving has been minimized and has been replaced with shelving that can be rolled to a new location as the shop floor layout is continually improved.

Have Machine Setup Times Been Reduced and Workload Been Leveled and Balanced?

Machine setups within manufacturing and assembly cells have been reduced to enable mixed-model scheduling (processing shorter runs of different products more frequently). Setup reduction efforts have separated internal and external activities and eliminated all adjustments (Shingo 1985). Production rates of assembly cells are managed by adjusting the number of workers within a cell and/or using additional cells as needed. Production planning and control issues are being addressed to balance and level workloads among cells.

Have Quality and Preventive Maintenance Programs Been Established?

A quality program has been implemented to provide the workforce with the tools and skills needed to enable continuous improvement. Workers assume responsibility for product quality. Workers are actively using statistical process control and experimental design techniques to learn how to control poor quality operations. Mistake-proofing mechanisms are being employed to ensure first-time quality (Nikkan Kogyo Shimbun 1988).

A preventive maintenance program has been established with machine operators doing simple regular maintenance activities. Maintenance procedures are documented and logs are kept of equipment maintenance history. The shop floor is operated at undercapacity conditions that ensure long operating life and allow worker housekeeping activities to minimize unscheduled downtime.

Has Mixed-Model Scheduling Been Established?

Production planning and control has implemented mixed-model scheduling in final assembly to permit the production of consistent product mixes over the planning time horizons. Material kanban has been implemented to pull fabrication and subassembly parts through the manufacturing system. Specific bottleneck and batch operations are rescheduled daily. As a result, in-process and finished goods inventories have been reduced permitting faster response to changes in customer demands.

Are Non-Value Activities Identified and Being Eliminated?

A program of vendor certification has been established, and incoming inspection is being phased out. Authorization and approval processes have been streamlined, eliminating most multiple review and approval activities. Layouts in office environments are redesigned reflective of frequent interactions among key design and manufacturing personnel. Auditing programs have been established to replace 100 percent review and approval processes.

Have Simple Automation Opportunities Been Exploited?

Machines that require constant operator attendance have been modified to permit single-cycle operation (that is, a self-feeding machine operation that completes an action when initiated by the operator—see Black 1991, 184–88). New machines purchased to replace existing equipment are all capable of single-cycle operation. Cells have been grouped to allow one operator to manage several machines operating from among various cells simultaneously. Some manufacturing cells have had automated material transfer equipment and simple robotics installed to free up operator effort and/or increase cell throughput.

Technology Applications That Enhance Agility

You have a strategy for implementing new technology effectively. You have laid all the groundwork and done the preliminary activities. Now you must select the right technology to help make your business more effective in rapidly responding to changing customer needs and market opportunities.

There are five fundamental ways to enhance design and production flexibility.

1. Reduce the product development time to market.

2. Reduce the product delivery time to the customer.

3. Improve workforce capabilities.

4. Enhance the flexibility of the production facilities.

5. Improve understanding and control of production processes.

Each will be discussed in more detail, with examples of technology that support these objectives.

Reduce the Product Development Time to Market

A major competitive advantage of the agile manufacturing enterprise is its ability to respond quickly to changing customer requirements. Typically, higher levels of customer responsiveness require drastic cycle-time reductions throughout the product development cycle. As shown in Figure 6.1, product development activities may include product definition, conceptual design, detailed design, prototyping, testing, producibility, and pilot production among others. Notice in the diagram that these activities overlap with one another. This is indicative of the iterative nature of the product development cycle. A product is initially defined by sales and marketing and then passed along to product design. The designers may have some questions about specific customer requirements and a negotiation between product design and definition ensues. As a product design becomes more concrete, prototypers become more involved. Difficulties with prototyping may require changes in both product design and product definition. Therefore, the process proceeds iteration by iteration. The application of technology in these activities can be important for satisfying customer demands for responsiveness and improved product performance.

CAD solid modeling is the foundation of agile product development technologies. This technology typically is used by product designers to

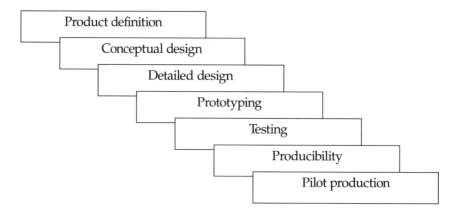

Figure 6.1. Typical product development cycle.

develop the geometric as well as nongeometric data about the manufactured product. Such data are then used in subsequent product development activities such as product prototyping, testing, producibility, and pilot production. As such, solid modeling technology becomes the hub of all product development activities, improving the efficiency of information flow and trimming product development cycle times.

By definition, a solid model is a computer-based, geometric representation of a bounded three-dimensional volume. Quite typically, solid models are composite assemblies of multiple basic graphical entities such as cones, cylinders, spheres, cubes, and so on. The volumes from these graphical entities are added and subtracted to one another to form the final product geometry. This is in contrast to wire frame models, which are simply sequences of line segments connected together to form two-dimensional (and sometimes three-dimensional) pictures.

Once completed, solid models include both graphical data about curves and surfaces as well as nongraphical data concerning relationships between graphical entities. For example, a solid model of a piston may not only provide the geometry of the part but also data for relating its length and diameter to the overall capacity of the engine in which it is to be installed. Most solid models are even capable of collecting information as diverse as material properties, cost, and manufacturing sequence. In contrast, wire frame models only contain the line segments used to define the product geometry.

Only solid modeling (as opposed to two-dimensional or three-dimensional wire frame modeling) can effectively take advantage of the increasing number of computerized tools becoming available to product developers. CAE tools can be used to evaluate mechanical, electrical, thermal, and flow characteristics of solid models using finite element analysis methods. In many cases, the same solid model can be used to evaluate the producibility of the product using *virtual manufacturing tools*. For instance, a CAD solid model of an injection-molded part can be used to directly generate a CAD model of its injection mold. This "virtual mold" can then be used to augment the mold design by simulating flow characteristics of the melted plastic during material injection. These options, as well as those for machining, die casting, and other production operations, are readily accessible through most CAD solid modeling software systems. Through the use of these virtual tools, product development problems can be caught much sooner in the product development cycle cutting down on the number of development cycle iterations and reducing product development time and cost.

Other CAE tools include the *design for manufacturability and assembly* (DFMA). DFMA tools apply axiomatic design principles to a product definition for the purpose of reducing the number of manufacturing process steps. For example, CAE modules exist that can be used to reduce the number and types of mechanical fasteners required within a product. Fewer mechanical fasteners improves the speed with which products can be assembled on the manufacturing floor and reduces the number of parts that must be located by the purchasing department. As such, DFMA can have a dramatic impact on the flexibility and efficiency of shop floor operations (see chapter 5 for more details).

Within the electronics industry, DFMA principles for modular design may be applied to both hardware and software. Many electronic products are now being built with embedded software to add product flexibility. Just as the hardware for many tailored electronic products are simply composite assemblies of smaller, standardized modules, so software is becoming more modularized. To prevent software development and configuration management from constraining the agile manufacturer, greater reliance is being placed on *real-time object-oriented software engineering*. As this field develops, electronic manufacturers will either develop object libraries in-house or rely on industry-specific libraries, probably provided by consortia or industry trade associations. Industry standards for object communication will need to be adopted to allow hardware components with embedded intelligence to communicate effectively.

Other uses of CAD solid models include the rapid fabrication of physical prototypes using *rapid prototyping systems*. Rapid prototyping systems include the use of both subtractive as well as additive fabrication equipment (Paul and Ruud 1996). Subtractive equipment is essentially conventional CNC machining equipment that can be used to rapidly fabricate part prototypes by removing material from a standard size workpiece. Additive equipment is newer technology that

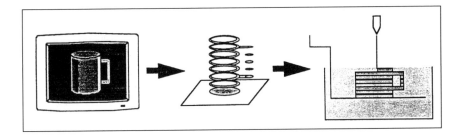

Figure 6.2. Process flow for additive rapid prototype system.

aims to build up part prototypes one thin layer at a time. A basic process flow diagram showing the steps involved with the additive rapid prototyping equipment is shown in Figure 6.2. The process starts with the development of a solid model on a CAD workstation. Next this solid model is dissected into many thin cross-sections on the order of several thousandths of an inch in thickness. These cross-sections, or slices, are then used to generate the control data needed to guide the delivery of energy to each layer of material.

Figure 6.3 shows a schematic of stereolithography (SLA), the first and most popular additive rapid prototyping technology available. SLA employs ultraviolet radiation in the form of a computer-controlled laser to selectively polymerize a light-sensitive polymer.

The advantage of additive systems over the more conventional subtractive systems is that they fabricate prototypes without the use of specialized tooling, thereby eliminating costly delays associated with ordering special fixtures and jigs. Another advantage of the additive systems involves the ability to prototype more geometrically complex parts. Subtractive technologies have the advantages of speed, accuracy, and the ability to fabricate prototypes in metal.

These inexpensive prototypes can be used to shorten cycle times in prototyping, testing, producibility, and pilot production. Manufacturers use rapid prototypes for various reasons, including design verification, bid requests, marketing demonstrations, and even functional testing. However, most rapid prototyping technologies are limited to fabricating prototypes in polymeric materials. Therefore, while the geometry of metal and ceramic parts can be verified using rapid prototyping

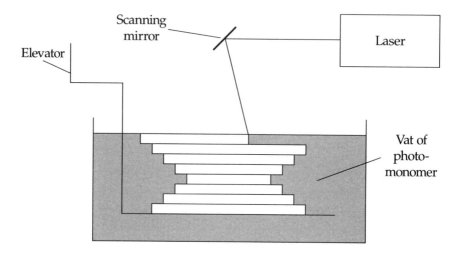

Figure 6.3. Schematic of stereolithography.

systems, the use of the polymer prototype in functional testing is limited.

More recently, progress has been made in using the polymer proto-types as patterns and prototype molds in secondary processes used to make real metal and ceramic parts. For example, rapid prototypes have been used extensively as patterns for sand, plaster, and investment casting processes in order to produce physical metal parts (Mueller 1992a; Mueller 1992b; Sarkis and Kennerknecht 1994, 291–300).

Much of this technology can be accessed through regional job shops and service bureaus, which exist expressly to meet the needs of smaller companies that cannot afford the capital cost of the rapid pro-totyping equipment. However, a requirement for working with most of these vendors is a CAD solid model.

For companies that do not have a CAD system or need to upgrade their product drawings from wire frame to solid model, it is generally better to start over in reconstructing the solid models on CAD work-stations. While this can seem to be a labor-intensive process, *computer-ized reverse engineering* (or shape-digitizing) technologies exist to help small manufacturers quickly generate CAD solid models from physical parts to reduce the effort needed for generating solid models of existing product lines. Techniques for implementing shape digitizing technologies include both contact and noncontact techniques (Bidanda, Narayanan, and Billo 1993). Contact techniques include the use of coordinate measuring machines, while noncontact techniques often employ lasers, ultrasonic equipment, or radio-frequency equipment in digitizing physical parts.

Reduce the Product Delivery Time to the Customer

For most manufacturing operations, product development and product delivery are separate cycles. Once a product has been developed and integrated into the production planning of a manufacturing facility, it can be ordered by a customer. Thus, reducing the time required to deliver an integral, quality product is also important to satisfying customer demands. More often customers expect responsiveness in addition to economy, quality, and value.

From the perspective of agile manufacturing, the most significant impact of reducing production cycle times is to reduce the dependence of production scheduling on forecasted demand. Many companies respond to customer demand for products from warehouses full of finished goods inventory. By reducing the cycle times involved with producing a product, finished goods inventories can be reduced resulting in cost savings. In addition, responsiveness to changes in customer demand are improved. Thus, the organization can move toward building only confirmed orders in quick response to their customers' needs.

Efforts can be made to implement production technology to reduce the product delivery time from order receipt to product shipment. *Manufacturing resource planning systems* can be used to share common information to permit several employees to process different parts of an order in parallel, rather than sequentially, thus reducing cycle time. For example, engineering bills of material can be used by purchasing to locate part vendors as well as by production control to specify inventory pick lists for shop floor consumption. These systems eliminate the rekeying of data and improve data reliability. More sophisticated *interorganizational information systems* can allow customers to provide self-service, enabling them to configure and place orders electronically and then directly query order status. When these systems are based on industry standards for sharing data, they are known as *electronic data interchange* (EDI). As customers demand more customized products, quick and accurate product pricing that reflects resource consumption becomes more critical. Software to provide *activity-based product costing* for different configurations of products can be an enabling technology to support quick and easy ordering of make-to-order items. Outside consultants can be sought to design and implement these types of manufacturing information systems to reduce manufacturing cycle times.

Improve Workforce Capability

The quality of an organization's employees—the scope and depth of their knowledge, their ability to solve problems, and their ability to learn and absorb new methods and technology—is a key aspect of becoming and *remaining* agile. Managing the process of continuing education and training is critical to the success of the agile organization (see chapter 9). For example, as previously noted, the benefits of CAD solid modeling technology are significant. However, very few active product design engineers actually know how to use CAD solid modeling technology effectively. While education and training issues cannot be solved by technology alone, several current and emerging technologies can play a supporting role in delivering quality training and education more quickly and cost-effectively.

The agile manufacturing organization has several needs concerning training and education. The cost of providing training needs to be controlled. The time and cost to develop and deliver new training and work instructions can become a limiting factor to organizational learning. Tools and decision aids that support frontline workers need to be produced more quickly and cheaply. For example, color coding can be an effective means of visual control on the shop floor. However, at some point product complexity can be too large to be accommodated by a limited number of colors. To overcome this, a remanufacturer of transmissions installed a product bar-coding system tied to an on-line technical documentation system to help operators rebuild many different types of transmissions in the same assembly cell (Hall and Anderson 1993). This system was developed in-house from readily available computer technology.

Interactive CD–ROM technology can provide enormous amounts of education and training in a decentralized manner. The cost of *teleconferencing* is likely to continue to decline and make remote problem solving and as-required education more common. Finally, object-oriented software engineering is providing useful tools for quickly developing multimedia educational software.

Enhance the Flexibility of the Facilities

Layout and equipment in a production facility have a profound impact on the agility of a manufacturing system. As such, relayout of an existing facility is usually a top priority in the transition to agile manufacturing. Often, significant improvements can be made immediately with modest investment, but near-term compromises often must be accepted due to inflexible characteristics of the building and equipment. Specific equipment may be difficult to relocate or not easily placed in several manufacturing cells due to overcapacity or other factors. This equipment is like a monument—often impressive, but frequently dedicated to an idea with limited relevance to current events. The long-term goal of the agile manufacturer is to have facilities and equipment that support agility and easy reconfiguration to take advantage of continuous improvement and changes in production requirements. Expensive, highly complicated pieces of equipment may actually reduce agility. For example, for the same price, a highly automated machining center has much less flexibility and capacity than multiple *single-cycle automatic machines.*

Facility flexibility can be achieved with a number of supporting technologies. Facilities can be equipped with *quick-disconnect utilities* (power, compressed air, venting, and so on) to minimize the time and expense of modifying current layouts. Where material handling cannot be eliminated, fixed lifting equipment can be minimized using *specially designed carts and guided vehicles.* Fixed walls and barriers should be minimized as well. In the near future, they may not be needed to control noise levels, having been replaced with equipment using *active noise suppression technology.* This technology will also make it easier to establish manufacturing cells without producing noise levels that endanger worker hearing and interfere with worker communication.

Often equipment is centralized because of environmental or occupational safety reasons. Changing the process technology can reduce or eliminate these concerns, making it easier to locate several smaller pieces of equipment into cells. For example, *modular spray cleaning equipment* and *powder painting* can eliminate the use of volatile organic chemicals that require elaborate recovery and control.

Where equipment cannot be decentralized, technology may play a useful role in increasing flexibility by reducing setup time and supporting mixed-model production. For example, Nissan has installed a *software-controlled fixturing system* to support robot welding of its car frames (Moskal 1992). Technology also can be used to make equipment more *ergonomically flexible* to support easy transfer of employees among different jobs.

Improve the Understanding/Control of Production Processes

Agile manufacturing is constrained by two key factors: (1) the ability to control production systems and (2) the limits of understanding the physical phenomena underlying production processes. New technology has a role to play in both areas.

As an organization moves toward agile production, it will likely implement a number of *visual control techniques* (Greif 1991). Many of these will be simple manual devices. For larger and more sophisticated needs, the organization is likely to implement a series of *electronic visual control mechanisms.* The most challenging aspects of this technology is to identify the various information flows and link them with the appropriate control mechanisms. *Simulation software* can be helpful in evaluating alternative production systems and guiding the effective implementation of visual control.

As mentioned elsewhere in this book, consistent product quality is a requirement of agile manufacturing systems to eliminate work stoppages and waste. Technology can be employed to assist in this pursuit in two ways. Where understanding of the underlying process physics in a production process is limited, improvements in process capability are constrained to *experimental design* and *statistical process control* techniques. As such, *automated data collection, analysis, and presentation* can be used effectively to assist in facilitating these techniques.

Where process physics are well understood, real–time data collection from *intelligent sensors* can be combined with *computer process models* to develop *integrated inspection systems* for improved process capability.

Agile Technology Make-or-Buy Decisions

When deciding to use agile manufacturing technology, several questions must be answered, including whether to buy the technology and develop the internal technical expertise or to outsource technical work to vendors. If a company makes the commitment to purchase technology, the company must determine which piece of equipment to buy. If the company outsources, it must decide which vendor to choose. In addition, the company must recognize and prepare for the effect that agile manufacturing technology will have on the organization. Adopting rapid prototyping equipment and solid modeling software provide good illustration of these issues.

As suggested elsewhere, answering these questions begins by understanding the strategic direction for the enterprise and identifying how the technology supports the strategic direction. Further, comparisons must be made between the application requirements and the technology capabilities. For example, in seeking to make use of rapid prototyping equipment, a manufacturer of engine blocks must realize that only three or four of the equipment vendors make machines large enough to fabricate engine blocks. Other requirements important in choosing the rapid prototyping technology may include dimensional tolerances, surface finish, build time, and material properties of the prototype part.

Finally, a comparison of costs and benefits can help make the final decision whether to buy or outsource. In all, the personnel, maintenance, training, and site preparation costs must be anticipated in order for the implementation to be a success. For example, when considering rapid prototyping systems, several costs must be factored into its procurement, including

- Equipment costs
- Support technology costs, including CAD systems, cleaning equipment, postcuring devices, material-handling equipment (for unstaffed subtractive fabrication), and so on

- Training costs
- Freight and installation costs, including any building modifications

Beyond procurement are the costs required for its operation, including

- Material costs, both raw materials and consumables
- Maintenance costs, including service contracts, preventive maintenance, equipment downtime, and so on
- Quality costs, including scrap
- Facility costs, including floor space, utilities, and so on
- Personnel costs

For small companies with tight budgets, the predominant issue in procuring agile manufacturing technology involves lowering initial costs. For example, consider the transition to solid modeling. Currently, the base version of ProEngineer™ (a state-of-the-art solid modeling software package capable of supporting all of the functionality described previously) can be acquired for around $30,000. At the low end, AutoCAD™ has recently released a solid modeling product for $10,000. The functionality of the low-end systems should be expected to be reduced with fewer provisions for parametric and constraint-driven design. In addition, ProEngineer™ has been on the market much longer and has a larger number of product extensions (for example, CAE modules) to offer. Both of these systems can be run on 486/66 MHz machines costing less than $2000.

For companies that have already made an investment in CAD technology, a major issue involves the protection of existing investments. Some packages offer a migration path to solid modeling. For example, for $500, AutoCAD™ users with traditional three-dimensional wire frame capabilities can purchase a solid modeling upgrade. As expected, this upgrade provides limited functionality and will not automatically translate previously constructed wire frame models to solid models. However, for the company tight on capital appropriations, this approach may offer a solution.

Another issue for these companies may involve the retraining of existing CAD personnel. Solid modeling is very different from wire frame modeling. For example, in the construction of solid models for mechanical parts, the requirements of solid modeling personnel would include the ability to understand the part as a machinist would. That is, the operator must see the product model as a block of material that has had material removed from it in various geometries. Through the union and intersection of various geometries, end parts are built. On the other hand, wire frame models are generally built by specifying surfaces and edges. However, experience has shown that longtime wire frame modelers have been able to adapt their skills to solid modeling with one week of training and less than 20 additional hours on the machine.

Experienced personnel are very important when acquiring agile manufacturing technology. For example, stereolithography machines have over 30 input variables that require fine adjustment in order to make good parts. In addition to experienced personnel, some estimates have suggested an average of one person per rapid protyping machine for simple part finishing (Wohlers 1991). Obviously, this can vary depending upon the machine and its application. Maintenance contracts for rapid prototyping systems can run between $5000 and $67,000 per year. Site preparation also can be a factor. Certain rapid prototyping systems use toxic resins that require high-capacity venting or cleaning solvents that must be changed frequently. Many systems require additional room for postprocessing, including cleaning and postcuring.

Prior to procurement of agile manufacturing technology, it is generally a good idea to test a particular technology by using the services of a job shop or service bureau. Over 60 rapid prototyping service bureaus exist in 20 states throughout the United States (*The Rapid Prototyping Directory* 1994). In addition to trying technology through a job shop or service bureau, many small and medium-sized companies prefer to outsource work over the cost of buying the technology. Here are some of the benefits associated with buying versus outsourcing (Burns 1993).

Outsource

- Save on capital, space, and personnel costs
- Immediate expertise available in the appropriate technology

- Freedom to use different technologies for different projects
- No risk of investing in the "wrong" technology

Buy

- Lower cost for high-volume usage
- Data and design security
- Immediate attention to priority projects
- No lead time or lag time delays in using technology
- Develop expertise in-house

When selecting a service bureau or job shop, several considerations must be made. First, does it have compatible CAD systems? For example, in preprocessing a CAD file for a CNC application, a significant portion of time can be spent simply getting the original CAD file into a recognizable format. Also, a file that has been converted from one CAD system format to another is much more likely to cause difficulties during CAM processing (for example, tool path generation). Many job shops have engineering services, which are important if experience in CAD solid modeling is lacking. It is important to anticipate the cost of CAD rendering if necessary, as it can be a sizeable portion of the overall cost.

Second, it is important to investigate whether the vendor has the type of technology needed for the application. Many job shops offer multiple, competing technologies and can aid in deciding which technology is best for the application. Once a technology has been selected, be sure that the vendor performing the work has adequate experience in using the technology. Ask for references if necessary.

For more information concerning the availability of job shops and service bureaus in a particular area, consult *The Rapid Prototyping Directory* (1994).

Technology Assistance Available to Manufacturers

Many levels of expertise exist for supporting the installation and/or use of appropriate manufacturing technologies. Regional job shops and service bureaus exist for outsourcing specific product development activities including solid model development, part prototyping, and prototype tool fabrication. Most technology vendors provide the expertise for installing equipment and training personnel in how to effectively use it.

Such product design and manufacturing services provided by commercial vendors can be expensive for small manufacturers. As a result, some government assistance is being organized in technical assistance centers, user facilities, and shared manufacturing facilities across the country. The National Institute of Standards and Technology (NIST) in Gaithersburg, Maryland, is an excellent source for locating technical assistance through its Manufacturing Extension Partnership (MEP) Program. Currently, there are more than 40 MEP centers across the country, which specialize in manufacturing technology assistance. Services range from technology planning to training to business consultation. A list of these MEP centers is provided in Appendix B in this book. For more information concerning MEP services and center locations, contact

> The Manufacturing Extension Partnership Office
> The National Institute of Standards and Technology
> Building 224, Room B115
> Gaithersburg, MD 20899-0001
> Telephone: 301-975-5020
> Fax: 301-963-6556
> E-mail: MEPinfo@micf.nist.gov

Because of the expense associated with the new agile manufacturing technologies, many user facilities are being developed to allow small and medium-sized businesses access to this technology for little or no cost. Sponsors for these user facilities range across industry, government, and academia.

An example of an agile manufacturing user facility exists at Oak Ridge National Laboratories (ORNL) in Oak Ridge, Tennessee. ORNL has made available to regional small businesses the pilot production facilities and concurrent engineering services and expertise to quickly prototype new product design concepts. Similar assistance programs are being pursued at many of the other Department of Energy (DOE),

Department of Defense (DoD), and National Aeronautics and Space Administration (NASA) federal laboratories, as well as within the NIST MEP program previously mentioned. More information regarding the manufacturing technology assistance services of the federal laboratories may be sought through the federal laboratories database. For more information concerning the federal laboratories database, contact

> Mid-Atlantic Technology Application Center
> Pittsburgh, PA
> 800-257-2725

Another source of information about technical assistance centers and user facilities at the state and local levels of government can be supplied by the National Association of Manufacturing Technology Assistance Centers (NAMTAC). NAMTAC can be contacted through

> NAMTAC Membership Services
> 119 Bird Building
> Western Carolina University
> Cullowhee, NC 28723
> 704-227-7059

In addition, the book *Partnerships* (Coburn and Berglund 1995) is a potential source of referrals to technology assistance available to small and midsized manufacturers.

References

Bidanda, B., V. Narayanan, and R. Billo. 1993. Reverse engineering and rapid prototyping. *Handbook of automation and manufacturing,* edited by R. C. Dorf and A. Kusiak. New York: John Wiley & Sons.

Black, J. T. 1991. *The design of the factory with a future.* New York: McGraw-Hill.

Burns, M. 1993. *Automated fabrication: Improving productivity in manufacturing.* Englewood Cliffs, N. J.: Prentice Hall.

Chew, W. B., D. Leonard-Barton, and R. E. Bohn. 1991. Beating Murphy's law. *Sloan Management Review* 32 (3): 5–16.

Coburn, C., and D. Berglund. 1995. *Partnerships: A compendium of state and federal cooperative technology partnerships.* Columbus, Ohio: Battelle Press.

de Treville, S. 1992. Time is money. *OR/MS Today* 19 (5): 30–34.

Greif, M. 1991. *The visual factory.* Portland, Oreg.: Productivity Press.

Hall, R. W. 1983. *Zero inventories.* Homewood, Ill.: Dow Jones-Irwin.

Hall, R. W., and J. W. Anderson. 1993. Enterprise manufacturing. *Target* 9 (5): 20–27.

Harmon, R. L., and L. D. Peterson. 1990. *Reinventing the factory: Productivity breakthroughs in manufacturing today.* New York: The Free Press.

Harrel, C., and K. Tumay. 1995. *Simulation made easy: A manager's guide.* Norcross, Ga.: Industrial Engineering & Management Press.

Moskal, B. S. 1992. Nissan heads "anywhere." *Industry Week* 241 (1): 19.

Mueller, T. 1992a. Stereolithography: A rapid way to prototype die cast parts. *Die Casting Engineer* 36 (3): 28–33.

———. 1992b. Using rapid prototyping techniques to prototype metal castings. *SAE Technical Paper* 92-1639, 1–5.

Nikkan Kogyo Shimbun. 1988. *Poka-yoke: Improving product quality by preventing defects.* Portland, Oreg.: Productivity Press.

Paul, B. K., and C. O. Ruud. 1996. Rapid prototyping and freeform fabrication. In *Integrated product, process, and enterprise design,* edited by B. Wang. London: Chapman & Hall, forthcoming.

The rapid prototyping directory. 1994. San Diego, Calif.: CAD/CAM Publishing.

Sarkis, B., and S. Kennerknecht. 1994. Rapid prototype casting: The fundamentals of producing functional metal parts from rapid prototyping models. *Proceedings of the Fifth International Conference on Rapid Prototyping.* Dayton, Ohio: University of Dayton Management Development Center.

Schonberger, R. J. 1986. *World class manufacturing: The lessons of simplicity applied.* New York: The Free Press.

Shingo, S. 1985. *A revolution in manufacturing: The SMED system.* Portland, Oreg.: Productivity Press.

Wohlers, T. T. 1991. The real cost of rapid prototyping. *Manufacturing Engineering* 107 (5): 77–79.

CHAPTER 7

Strategic Direction
Russ E. Rhoads

As an organization makes the transition to agility, its position relative to its markets, customers, and competitors changes dramatically. The old strategic equation no longer applies; the firm takes on a new strategic identity. An agile firm competes on a different basis. New practices are required for creating and implementing strategies to capitalize on this new position. This chapter describes the processes and systems that can be used to set and implement strategic direction in the agile firm.

There are three key strategic practices required to build and maintain an agile manufacturing firm. These include

- *Strategic thinking* to develop and maintain a continual focus on the long-term success of the firm

- *Strategic learning* to continually assess strategies and adjust them as needed to remain successful in a dynamic environment

- *Strategic partnering* to capitalize on the size-related advantages of the small to medium-size business and mitigate the disadvantages

Each of these practices is discussed in this chapter and mechanisms are described for implementing the practices in a small to medium-sized manufacturing firm. Prior to that discussion, some basic terms and concepts are presented to provide the common language needed for the remainder of the discussion.

Key Concepts

A *strategy* is the approach taken by a firm to achieve and sustain competitive advantage. Strategies implement the guiding vision described in chapter 9. The vision describes the ultimate future for the firm. Strategies represent the actions that will propel the firm along the path toward achieving the vision. To establish a successful strategy, the decision makers in the firm must examine long-term markets, assess the firm's strengths and weaknesses, anticipate likely actions of competitors, determine directions key customers will go, and understand the capabilities that will be available through technology to develop new products, improve existing products, or enable improvements in production or support processes. They must use this information to determine how to achieve the best match between the opportunities and risks in the marketplace and the resources and aspirations of the firm. This is a complicated task involving many factors that can only be approximated.

The potential difficulties associated with establishing a winning strategy are illustrated in Figure 7.1. The key strategic forces that shape the future of the firm are illustrated along the left side of the figure. These forces include the following.

• *Market factors* are most directly reflected in the needs and future directions of current and potential customers of the firm. These factors also include possible actions of competitors and directions of key suppliers. It may be necessary to examine the forces driving demand to accurately interpret market information. These forces might include demographics, pressing social problems, or trends in attitudes or values.

• *Business factors* are driven by the state of the economy. These factors include growth rates in the economic sectors in which the firm does business, cost of capital, and demand for labor skills the firm needs. Also included are the legal, regulatory, and political climates in which the firm will operate.

• *Technology factors* include technology developments that could enable the firm to improve its products or the processes used to produce these products. Technology developments can also give competitors

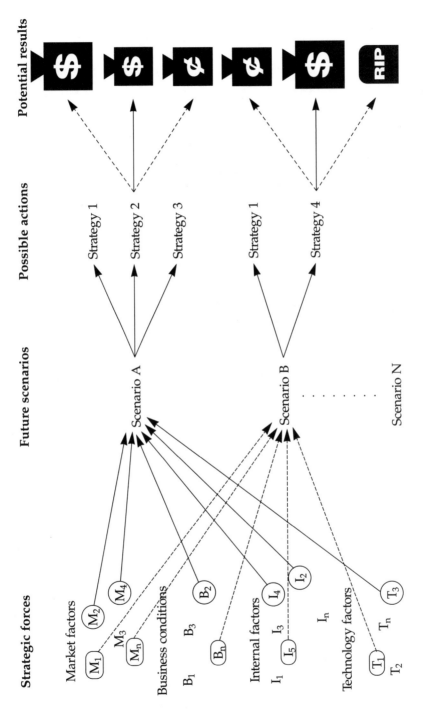

Figure 7.1. To chart a successful path to the future, the agile organization must deal with many complexities and uncertainties.

193

potential advantages and impact the demand for products and services of the firm or its major customers.

• *Internal factors* reflect what is going on inside the company. They include how the company views itself—its values, competencies, and vision of its future. They also include the condition of plant and equipment; skills of the workforce; financial position; and availability of intellectual property, patents, or licenses that could be used to build and sustain a competitive advantage.

These factors are interrelated and combine in complicated ways to shape the future in which the company will do business. From the present, it is not possible to predict with certainty what will happen in the future. Many futures could develop. This is illustrated in Figure 7.1 by the multiple scenarios that could unfold. Two possible scenarios are shown, although there are a large number of possibilities.

In response to these scenarios, the firm could follow a number of possible strategies. In Figure 7.1, there are three possible strategies the firm could pursue in scenario A and two major strategies in scenario B. In this illustration, strategy 1 would apply in both scenarios. Each of the strategies the firm might pursue could have several possible outcomes, depending on how future events unfold. The solid line from the strategy shows the most likely outcome. Other outcomes could result in better or worse financial returns for the firm, possibly even threatening the survival of the business.

Figure 7.1 shows the complicated array of future possibilities that must be dealt with to set an effective future direction for the firm. There is, of course, no way to accurately predict the future. However, methods are available to chart a successful course through this complicated array of factors. The basic process of setting and implementing strategy for the firm is illustrated in Figure 7.2, and includes

• Understanding the strategic forces at work both internally and externally to the firm

• Identifying potential winning strategies for the firm

• Selecting and implementing the best strategy

• Monitoring to detect significant changes in the key factors that drove the selection of the best strategy

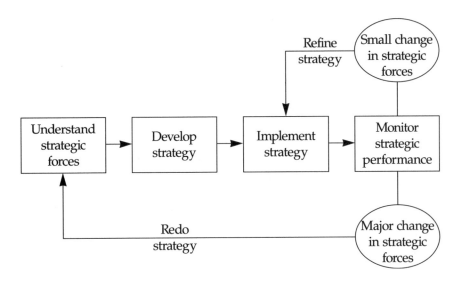

Figure 7.2. The basic process for setting strategic direction.

- Refining or redoing the strategy in response to these changing forces

The remainder of this chapter describes how agile strategic practices can be used to carry out this process successfully in a small to medium-sized firm.

Strategic Thinking

Strategic thinking is the foundation of strategic agility. The firm that has decided to "go agile" has done so because it has examined the factors influencing its future and concluded that long-term success is based on its ability to respond flexibly and quickly to changing customer needs. To achieve strategic agility, the firm must take the additional steps required to make this strategic view an everyday part of the business, not just an event that takes place from time to time. Rather than pausing occasionally to look at the future, the agile firm strives to fill the organization with strategic thinkers who always consider long-term implications when they make day-to-day decisions.

Strategic thinkers are simultaneously farsighted and nearsighted. They know that sustaining a positive cash flow and meeting delivery schedules are critical for near-term viability. They also recognize that increasing customer loyalty, maintaining a stable workforce, and increasing the competency of the firm are essential for long-term success. They understand that sometimes it is necessary to make trade-offs among these important factors. They realize that to give themselves the opportunity to make good trade-off decisions, they must conscientiously consider long-term implications in their day-to-day actions.

An organization made up of strategic thinkers is able to find and maintain the optimum balance between near-term gains and long-term success. Finding the optimum trade-off between near-term and long-term benefits requires people at all levels in the organization to make decisions in new ways. For example, consider a situation in which a key customer asks for a product modification that requires a capability not currently available in the firm. Taking a short-term perspective, the firm might subcontract the portion of the product it couldn't produce with current capabilities. This would avoid some investments in capital equipment and staff training, maintaining current profit margins. A strategic perspective would bring other considerations into the decision. From a longer-term view, the investment might be seen as an opportunity to broaden the company's product lines, better position the firm for technological changes occurring in its industry, and build even stronger ties to a key customer. From this perspective, the near-term reductions in profits may be more than offset by the long-term gains. Strategic thinking is required to make the best decisions in these kinds of situations.

The shift to strategic thinking must begin with the management in the organization. Managers must recognize the need for and the value of looking at the business from this different perspective before they can initiate the activities that make strategic thinking the "ordinary" way of doing things. In a small to medium-sized firm, this shift may be particularly difficult for the senior manager. Often the senior manager is the one who started the company. This person typically has an entre-preneurial bent. He or she is not only comfortable with "flying by the seat of the pants," but has been very successful operating this way. Considering strategic implications in day-to-day activities and becoming more deliberate about developing and implementing strategic direction will often be a major change for the senior manager. (This is not intended to suggest that senior managers abandon the management techniques that have helped them to be successful—only that they enhance their effectiveness by adding some tools to their repertoire.)

There is no magic potion to help managers make this transition. They must make this shift because they believe it is the right thing to do. The members of the management team can reinforce each other to accelerate the transition. They can put strategic implications on the agenda for all their formal meetings. They can agree to always end their informal discussions with a question like, "Are there long-term implications we have overlooked?" They can also remind one another when they slip back into old habits. Simple code words are often useful for this. The manager doing the reminding need only say the code word (for example, *strategy*) to remind his or her colleague. This can take the accusation out of the reminder and can even be fun if the team is somewhat creative in coming up with the code word. Formal mechanisms also should be used to facilitate the transition. The need for strategic thinking should be written into the performance expectations for each of the members of the management team, including the senior manager. It should be included in performance reviews and considered in salary actions, promotions, and decisions about employment benefits.

The management team has three kinds of tools available to help the rest of the staff make the transition to strategic thinking—communications, modeling the expected behavior, and performance measures. Patience and positive reinforcement will also be required, as it takes time and effort to change the vantage point from which the employees in the firm view the business. If the employees are to become strategic thinkers, they must first understand the strategic direction of the organization. Senior managers must spend time and effort communicating key strategies to all employees. Managers can help accelerate this change to a strategic perspective by modeling the shift. A simple way to do this is by explicitly bringing up strategic implications in day-to-day business discussions and explaining decisions in terms of longer-range implications. Many of the devices the management team used to make the transition to strategic thinking also can be used to help the rest of the staff establish good habits. For example, strategic implications can be placed on the agenda for every meeting.

Chapter 8 discusses one of the basic rules of organizational behavior: You get what you measure. This rule can be put to good use in helping people make the shift to becoming strategic thinkers. This is accomplished when senior managers establish good performance measures for strategic objectives and make information on progress toward these objectives readily available to everyone. For example, a firm that is being driven by market forces to "go agile" might establish a strategic objective to reduce by 50 percent the cycle time between expression of customer demand and delivery of the final product. An order tracking system might be set up to facilitate this improvement. Information from the tracking system could be used to produce regular reports on cycle time for this customer order fulfillment business process. The reports might be displayed on the wall in the shop floor, published in the company newsletter, or posted in the lunchroom so they are readily visible to all employees. This would encourage all employees to look for ways to help meet this key long-range objective as they go about their daily activities. The rewards and recognition system must also be aligned with the importance of strategic thinking. Promotions, salary actions, and formal and informal recognition activities should reward people who demonstrate strategic thinking.

With some effort and attention, strategic thinking will permeate all levels of the organization. The result is an organization whose members are moving together toward the key objectives that will ensure future success. It also creates the cohesiveness needed to change direction when strategic forces require modification to the objectives. An organization filled with people who make future considerations a part of their day-to-day activities will quickly shift to the new goals. This ability to change course rapidly when needed is a key element of strategic agility.

Strategic thinking is the foundation of strategic agility, but strategic learning is the core. Most of the real work of developing and implementing strategic direction in the agile firm is done through strategic learning processes. The bulk of the remainder of this chapter is devoted to the discussion of strategic learning.

Strategic Learning Approach to Creating Strategy

Strategy is about the future, but it is developed by examining the past. Therefore, it is an inherently uncertain process. Strategic learning is a practice that can be used to create successful strategies in a way that accommodates these uncertainties. Much of the literature on strategy development and strategic planning treats the development of the key strategies of a firm as an event. Strategic planners provide a great deal of information about the internal and external environments to senior managers who have some sort of "aha" about directions in which to move the firm to ensure future success. The flaw in this approach is that strategy setting is not a single event, it is a continuous process.

The strategic learning approach recognizes the continuous nature of strategy development and implementation, and treats it as a dynamic process. The key strategies of the firm are adjusted in response to changing conditions in the marketplace or new realities within the firm. A strategic learning approach permits the agile firm to balance necessary changes in strategy with "staying the course" when changes are not needed. This staying the course (often referred to in the literature as *strategic intent*) (Hamel and Prahalad 1989) is necessary to actually implement the strategy once it has been developed.

Figure 7.3 provides an overview of the strategic learning approach to strategy development. The remainder of the discussion in this section is organized around the major steps shown in the figure.

Establishing the Strategy Team

Development of strategy through a strategic learning approach begins with a commitment of time and resources to the task of strategy development. The small to medium-sized firm cannot afford the full-time strategic planning staff employed by larger firms to carry out these

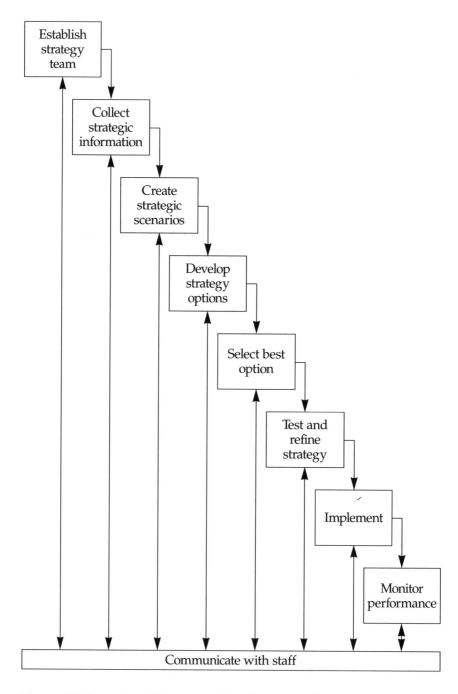

Figure 7.3. Overview of the strategic learning approach to setting strategy.

activities. However, this does not make the task any less critical to success. The first step is to designate someone or a small group of people as "strategists." In most cases, the senior manager will lead the strategy team. Often this is the founder, who knows the company inside and out.

It is important to note that the senior manager may have to alter his or her job to take on the strategic planning role. This is one of the key aspects of the changing role of managers in an agile organization, discussed in chapter 10. The senior manager may have to take a smaller role in day-to-day operations and devote more time and energy to long-term concerns. This shift makes sense for several reasons. First of all, it gives strategic planning necessary resources. Secondly, it begins a change that ultimately will be required. As the firm continues to grow, the senior manager must learn to delegate operational responsibility, or he or she becomes an obstacle to future success. This shift is often difficult to achieve. Devoting the energy of the senior manager to the most important task in the firm—developing and implementing the strategies that will ensure future success—should help with the change.

The strategy team will usually include a financial specialist, a production specialist, and a marketing specialist. A staff member from the product development group might also be included. People are assigned to the team based on their knowledge of the firm and its external environment, their ability to think strategically and creatively, their ability to collect and analyze strategic data, and their ability to think about the firm holistically (that is, not just from the perspective of their current area of responsibility). The fiscal realities of a small to medium-sized firm dictate that the members of the strategy team serve on a part-time basis. There will be short periods when they need to devote full time to these activities, but they will still carry out most of their regular responsibilities.

Generally, the senior manager will be the strategist or decision maker, and the remainder of the team will function more as analysts and advisers. The team will collect information, develop an understanding of the relationships among key strategic factors, delineate and test assumptions, and help to generate and evaluate alternatives. The strategist will evaluate information, synthesize factual information with intuitive insights, create novel options, and make final decisions.

Collecting Strategic Information

Unlike traditional strategic planning, information collection is not confined to the strategy team. An outcome of the switch to strategic thinking, discussed in the previous section, will be that all employees become sources of information that could be important to strategic decisions. Staff members who interface directly with customers provide an excellent source of information about customer needs and plans. Employees working in areas such as production usually have a good understanding of internal conditions and may have good information on emerging production technologies. Staff members in finance and accounting understand the fiscal position of the company and often will have good information on general business conditions. Staff members in procurement and other portions of the organization deal with suppliers on a regular basis. They should have good insights on actions key suppliers are likely to take. Some of the information staff provide to the strategy-setting process may come from unlikely sources. People have networks of friends that may have applicable information. They may have hobbies that provide them with knowledge about technology developments in applicable areas. There is often more and better information available than would appear at first glance. The challenge is to access it efficiently in the strategy-setting process. Figure 7.4 lists additional mechanisms that can be used to stay current on key strategic factors.

In a small firm, simple mechanisms can be set up to capture information important to the strategic planning process. Employees usually work closely with each other and with the senior manager, so this information can be passed along in the course of day-to-day activities. In a medium-sized firm, some form of internal network will usually be needed. This can be as simple as directing each department head to identify sources of strategic information in his or her organization, collect the information, and pass it on to the strategy team. Special informal meetings could be held to collect the information, perhaps as the preliminary event to a social get-together. If the company has a computer network in place, electronic bulletin boards could be used. An employee who encounters a piece of strategic information could simply post it to the bulletin board. If quality circles, or some equivalent, are

There are several areas of strategic information collection that represent special challenges for the small to medium-sized firms. These firms usually cannot afford to employ specialists to keep current on these critical kinds of information. Mechanisms that can be used to stay current on technology, market forces, economic trends, and other key strategic factors include

- Participate in consortia such as the National Center for Manufacturing Sciences (NCMS). NCMS is a large consortium of manufacturing companies that collects and disseminates information, supports the development of advanced manufacturing technologies, and fosters rapid transfer of these technologies into member companies.

- Take advantage of government programs designed to help small to medium-sized manufacturing companies to utilize advanced manufacturing technologies. (See chapter 6 for more details.)

- Attend meetings of technical societies. National meetings collect a great deal of technology information in one location. Many have extensive vendor trade shows in conjunction with the meeting. Local chapters of technical societies also can be a useful source of information. These chapters usually hold regular informational meetings and provide opportunities to network with people with special expertise.

- Work with vendors. Vendors can provide useful information about where technology is going and can provide ideas on how it might fit into your business. They will usually provide this information free in the hopes that you will buy their product.

- Develop a close working relationship with your banker(s). They can be a useful source of information on economic conditions, cost of capital, and expected changes in other key business conditions.

- Utilize consultants. They can provide special expertise and tailor it to the needs of your business. It takes some effort to sort out the best sources of this kind of assistance. Check out potential consultants as you would a key employee. Find someone who wants to understand your particular situation and then help you pick the tools that will be most helpful to your business.

- Develop a long-term relationship with a research and development (R&D) organization. The R&D partner can be an excellent source of information on technology developments and other factors important to the strategic decision-making process.

Figure 7.4. Acquiring strategic information.

used in the firm, strategic information collection could be made a standard part of the quality circle meetings.

An important feature of these information collection mechanisms is that they help to reinforce the concept of strategic thinking. If people are constantly aware of and looking for information that may have strategic implications for the company, they are thinking in strategic terms.

To make best use of the network of employees collecting strategic information, the strategy team must be in regular communication with staff throughout the organization. Figure 7.3 illustrates the importance of making this communication a dialogue that takes place throughout the strategic planning process. Staff members provide the strategy team with information they have collected, and the team issues progress reports. The progress reports reinforce the strategic thinking practice by exposing the staff to details of the strategy-setting process. Two-way communications facilitate implementation of the selected strategy by ensuring that a broad segment of staff understand the strategy and why it was selected. Regular communications should trigger ideas from staff members for creative strategy options and permit them to identify other information they have or can access that would be beneficial to the strategy-setting process. Involving staff directly in the strategy-setting process also reinforces employee empowerment, discussed in chapter 10.

Creating Strategic Scenarios

The information collection activities will generate a variety of information that might have relevance to the strategic direction of the firm. The strategy team must organize and process this information to identify events, trends, or issues that could be significant to the future success of the firm. A useful tool for sifting the strategically significant from the possibly relevant information is to create *strategic scenarios.* The concept of strategic scenarios was illustrated in Figure 7.1 and an example scenario is presented in Table 7.1. Scenarios are nothing more than a collection of plausible versions of what could happen in the future. They are based on a synthesis of the available information about what is happening in the market; what key customers are doing; what is happening with key suppliers; trends in business conditions and politics;

Table 7.1. Example strategic scenario.

Business conditions	• Economic growth remains low in the sector in which our firm operates. • Federal Reserve pushes prime rate to 10 percent to control inflation and strengthen the dollar on world markets.
Technology factors	• Power of desktop computer workstations continues to rise while prices continue to fall. • Fiber-optic network installation continues to accelerate. T_1 telecommunications capabilities are widely available within two years.
Market factors	• Demand for our key customer's products stagnates because of high interest rates.
Internal factors	• Up to 6 percent of revenues are available for investment. Larger investments not feasible.
Outcomes	1. Key customer adopts computer-integrated manufacturing (CIM) to increase market share by reducing costs and decreasing time to meet customer demand. 2. Key customer decides to deemphasize this product line and diversify into products in markets that are less sensitive to cost of capital.
Implications	1. Our key customer requires us to invest in CIM and link our production processes electronically with its as a condition for continuing our current business relationship. 2. Demand for one of our major products decreases by 50 percent. Overall revenues decline by 30 percent.
Critical success factors	1. Establish full electronic data interchange capabilities with our key client within three years. 2. Increase market share with other customers by reducing costs for this product line by 25 percent and reducing the time it takes to fill customer orders by 35 percent within 18 months.
Strategy options	1. Acquire a small computer specialty business to provide in-house expertise in electronic data interchange. 2. Implement just-in-time practices in the production system and reengineer the customer order fulfillment process.

what potential competitors are doing; and what is happening in technology areas potentially useful to the firm to enhance products, develop new products, or improve production or support processes.

Scenario-based planning provides a useful framework for considering the future without getting hung up on the uncertainties (and giving up on planning for the future) or ignoring them (and hanging on to your current assumptions about what is going to happen). It is a good planning tool for the agile firm because it recognizes that the firm must be able to deal with a range of future possibilities. The strategic decisions of the firm then focus around selecting what range of future possibilities the firm must prepare for and choosing the best ways to prepare. There are many ways to go about the scenario development process. Some involve elaborate methodologies to develop scenarios and software programs to track alternative scenarios and associated probabilities. Schwartz (1991) emphasizes that the real power in the scenario approach is in the perspective provided by viewing the future through scenarios. The small to medium-size firm can take advantage of this perspective while applying the relatively straightforward scenario development approach described here.

The strategy team initiates the scenario development activities by organizing the strategic information into major categories. The categories listed in the strategic forces portion of Figure 7.1 would be a good starting point for this effort. Subcategories might be useful to help organize the information, but the goal is to put similar information together to facilitate the development of the scenarios, not to create an elegant taxonomy of strategic data. Some example subcategories might include

- Market forces—customer directions, supplier directions, competitor directions, demographic and social trends
- Business conditions—financial and economic conditions, labor markets, regulatory and political directions
- Internal factors—financial position, workforce, plant and equipment, intellectual property
- Technology factors—product technologies, process technologies, technology-driven demand factors

Once the information is organized, the strategy team can identify scenarios by selecting combinations of the market factors, technology factors, business conditions, and internal factors. The strategy team will have to exercise considerable judgment in this part of the process, as there are a very large number of possible combinations. Fortunately, because the factors are interrelated, only some of the possible combinations represent realistic future scenarios. Nonsensical combinations are not considered.

It is not necessary to try to develop an exhaustive list of future scenarios. The purpose of the scenarios is to provide a framework for considering future possibilities. Representative scenarios are usually created. A representative scenario covers a portion of the spectrum of future possibilities. It can be thought of as a major path into the future. Many variations on the theme are possible, but they would produce similar implications for the firm. A set of four to six representative scenarios is usually sufficient to cover the range of possibilities. The team will probably want to identify a larger list of scenarios as a starting point, then narrow it down to the best set of representative scenarios. It is often useful to include some best-case/worse-case (or optimistic/pessimistic) scenarios in the set of representative scenarios. This will help to ensure that a complete range of future possibilities has been considered. A status quo or "current trends continue" scenario is often useful as well.

Each scenario describes a possible sequence of future events. To provide a concrete basis for strategy development, the strategy team must determine how the events and trends in a specific scenario could affect the firm. Effects are assessed by defining potential *outcomes* from each scenario and evaluating the *implications* of each outcome for the firm. Outcomes are developed by answering the question, "If this set of conditions were to occur, what would be the result?" Implications are developed by answering the question "If this outcome occurs, what effect will it have on our company?" There could be several possible outcomes from a given scenario, as the "players" in the scenario must make decisions as the scenario unfolds. Two possible outcomes are provided for the scenario presented in Table 7.1. Implications of each outcome are also described. Schwartz (1991) recommends the concept of plot lines to help describe scenarios and identify outcomes and implications. Standard plot lines can be constructed by examining the winners and losers in a scenario, by looking at who is challenged and what their response is likely to be, and by describing logical evolutions from current trends. Unconventional plot lines can be developed by postulating revolutionary changes, examining cyclic phenomena, or looking at the possible impacts of "mavericks."

The set of four to six representative scenarios with associated outcomes and implications is the jumping-off point for the actual strategy-setting activities. This set of scenarios provides the strategy team with a realistic look at what *could* happen in the future. It is not feasible for the firm to prepare for all the future possibilities represented by the scenarios. The team must choose a subset of scenarios containing the range of future events it will actually plan for. Strategies are developed to maximize the success of the firm over the range of possibilities represented by this subset. We will refer to this subset of scenarios as the *planning baseline* for the firm. Selection of the scenarios in the planning baseline is the second key decision the strategist must make.

Selecting the members of the strategy team is the first. This decision requires considerable judgment. The decision will usually be based on an assessment of the likelihood of the various scenarios actually occurring, the magnitude of the implications (either good or bad) for the firm, and the degree to which actions by the firm can impact the outcome. The strategist is in essence making an assessment of the potential risks and benefits to the firm and the degree to which the firm can take actions that produce positive outcomes. Much of this assessment will be done intuitively, as there are few hard facts upon which to base this kind of decision. The scenario framework is useful for thinking through the possibilities, but there is no substitute for good judgment in the decision-making process.

Developing Strategy Options

The real work of the strategic planning process is to develop strategy options for each of the scenarios and associated outcomes in the planning baseline. An effective way to initiate this activity is to translate the outcomes and implications into *critical success factors* for the firm.

Critical success factors are the things the company *must* accomplish to be successful if this particular outcome occurs. To be most useful, critical success factors should specify *what* has to be accomplished and *when* it needs to be completed. The what should be quantitative if at all possible (that is, it should include a *how much*). The most useful critical success factors will have the characteristics of the customer-centered performance measures described in chapter 8. Examples of critical success factors are presented in Table 7.1.

The strategy team will usually develop several strategy options for each scenario and associated outcome in the planning baseline. Options are developed by answering the question "How will we achieve the critical success factors?" For the company making the transition to agile practices, many of the strategies will be based on agility. Example strategy options are provided for the scenario in Table 7.1. The step that translates critical success factors into strategy options is the most challenging and important part of the strategy-setting process. A recipe for success in this portion of the process will include equal parts perspiration, information, and inspiration.

• *Perspiration* comes from a strategy team that believes in the importance of what it is doing and is committed to building the strongest possible future for the firm.

• Much of the *information* that is needed to identify strategy options will be generated during the information-collection step of the strategy-setting process. For example, new technologies, management methods, or organizational concepts can enable the firm to move in new directions. These strategic enablers can permit strategies that rely on improving current products or developing new ones. They can also permit the firm to reduce its costs or improve its level of service to its customers by changing the way it produces its products or runs its management and production support systems. Information on these kinds of enablers will be collected during the normal course of a good strategic information collection process. Other kinds of information may be collected specifically to support the identification of good strategic concepts. One particularly useful way to generate good ideas is by learning from what other successful firms do. This does not mean copying others, but examining what they do that makes them successful, then translating that to your firm. You can learn useful lessons from successful companies across a broad range of businesses. You need not be confined to companies in businesses similar to your own. This benchmarking approach is discussed in more detail in chapter 3.

• The *inspiration* needed to develop superior strategy options can be significantly enhanced by adopting the right frame of reference. This frame of reference will have several key characteristics. (1) *Focus on the customer:* Successful strategies will be customer-centered rather than competitor-centered. (2) *Creativity:* Good processes can help you collect the right kinds of information and consider it in the right order, but there is no real substitute for a good idea. (3) *Willingness to challenge traditional ways of thinking:* One of the hallmarks of an agile firm is that it is eager to consider new ideas and try out the ones that fit with its vision and values. (4) *Courage and risk taking:* The best strategies will balance the magnitude of the possible return with the potential risks. An element of courage will be required to take the best path, rather than the safe path.

The strategy team will want to pay special attention to "overarching" strategies that address several critical success factors. For the example provided in Table 7.1, the company might establish a strategic partnership with an R&D firm. This would provide access to the special technical expertise needed to achieve both critical success factors. (Strategic partnerships are discussed in more detail at the end of this chapter.)

There are a number of techniques that can used to encourage "unconventional" thinking during the option identification process. Brainstorming is a good way to generate lots of ideas. An exercise that maps the critical success factors against the available strategic enablers (new production or product technologies, better business practices, organizational or management system improvements, and so on) also can be useful. This will permit the team to think about how the enablers might be best used either individually or in combinations to achieve the critical success factors. The techniques are helpful, but as in most of the critical portions of the strategy-setting process, it really boils down to hard work and creative thinking.

There are a number of potential benefits associated with involving staff in addition to the strategy team in generating strategy options. Broad involvement reinforces the concept of strategic thinking in the firm and taps into the knowledge and skills of the staff to generate creative options. Strategic thinking is reinforced by exposing staff to the strategy development process. Strategy implementation also will be easier, because the strategies that are selected are well understood by the staff. Involvement also helps staff members to be more effective in identifying information with potential strategic significance. Staff involvement can be accomplished as part of the normal communications process or as special events. For example, several brainstorming sessions could be held, involving staff from different parts of the organization. The team could also communicate the critical success factors to staff members in a letter or informal information meeting and ask them to contribute their ideas.

Selecting the Best Strategy Options

The approaches discussed in the previous section will generate a number of strategy options for each of the several possible outcomes from each scenario in the planning baseline. The objective is to generate many options to encourage creative thinking and improve the chances that the superior options will be identified. This shopping list of options must be evaluated to determine which offer the best possibility of future success. The potential costs, risks, and benefits of the strategy options are the basis for this evaluation. Factors typically used in this assessment include the likelihood of achieving the critical success factors; the amount and timing of the investment required; the time required to fully implement the strategy; anticipated impact on revenues (including possible near-term downturns as the changes associated with the new strategy are implemented); when results will start to be seen; the range of scenarios covered by the strategy; and the impact on the staff.

If a large number of strategy options has been generated (for example in brainstorming sessions), the evaluation will usually proceed in two stages. In the first stage, some simple techniques will be used to screen the options down to a smaller set (perhaps half a dozen or so) that will be evaluated in more detail. The screening approach usually starts with feasibility testing. This is simply taking a critical look at each option using questions such as

- Can it achieve the performance targets contained in the critical success factors?
- Can it produce the required results in time?
- Can it be accomplished with the available investment resources?
- Is the option consistent with our organization's values and culture?

If the answer to any of these questions is no, the option is declared infeasible and removed from further consideration.

Additional screening may be needed if a relatively large number of feasible options remain. Feasible options can be screened by ranking the options and picking the top-ranked few for more detailed evaluation. A preliminary characterization of the option provides the information needed for the ranking process. Characterization starts with a preliminary description of how the strategy would be implemented—identifying the major steps in the implementation and estimating about how long they would take. This description usually relies on the judgment of the strategy team members, perhaps after consultation with other staff having expertise relevant to this option. The team would then use readily available information and rough approximation types of analysis to assess the performance of the option relative to each of the evaluation factors listed in the previous paragraph. These types of preliminary evaluations usually are sufficient to identify the set of "superior" options that will be considered more carefully before selecting the "best" option.

The best option can be identified using a more thorough version of the ranking process described in the previous paragraph. At this stage, the team will develop a more detailed characterization of how and when the option would be implemented and perform more careful analyses to estimate capital and operating costs, revenue streams, and impact on other critical performance measures such as cycle time or quality. It may be necessary to perform estimates for each of the scenarios in the planning baseline, because the results are likely to be scenario-dependent. This will enable the team to determine how robust the strategy is. A robust strategy will perform well over a broad range of scenarios.

Selecting the best strategy is an important decision for the firm. It is worth spending some time and effort to do some analysis and sort through the options carefully. It is important to remember that the analysis is a tool, not an end in itself. A common pitfall is to get stuck in the analysis—putting off decisions to collect more data or crunch more numbers. Part of the antidote to this problem is the results focus described in chapter 10. It is equally important to not lose track of the uncertainties associated with the results of the analysis. The associated pitfall is to make decisions based on small differences between one option and another, forgetting that the results of the analysis are only approximations. This pitfall can be avoided by expressing the results of the analysis as ranges, keeping track of the uncertainties explicitly. For example, the purchase cost of new equipment might be expressed as ranging from $10 million to $20 million with a best estimate of $17 million.

A number of approaches can be used to develop the estimates needed for the evaluation. It may be useful to get more benchmarking information, using the experience of other firms to guide the estimates. Some simple models may also be used to estimate revenue streams and impact on key financial factors. Consultants could be used to help make some estimates if special expertise is needed that is not available within the company. For example, your accounting firm might help construct an activity-based cost model of the company or a university professor might help estimate the purchase and implementation costs of a new production technology. A management consultant could help describe the steps needed to implement a new management practice and estimate the implementation cost and the expected benefits.

Another useful approach for evaluating options is to test the current strategies of the firm against the scenarios in the planning baseline. This will determine if conditions are changing sufficiently to warrant a shift in strategy. A useful framework for doing this is to list the assumptions that the company currently holds about its markets, business conditions, itself, and the impacts of technology on its business

(Drucker 1994.) These assumptions represent what the firm *believes* to be true about itself and the external environment in which it operates. They are the basis for day-to-day decisions that mangers and staff make and usually represent an accurate version of the real strategy the company is following. Comparing these assumptions to the scenarios will quickly identify where the strategy of the firm needs to be adjusted. Strategy options that address these areas would be strong candidates for detailed evaluation.

To select the best strategy, the decision maker will have to apply intuition and judgment along with the results of the evaluations. A major advantage of the learning approach to strategy development is that it does not require that a final decision be made at this point in the process. The evaluation only needs to pick the leading candidate. The front-runner is further evaluated by testing before the organization commits itself to the new strategy.

Testing and Refining the Strategy

Risks associated with major strategic moves can be substantially reduced through testing. The tests make it possible to validate the strategy (or invalidate it, as the case may be) in much less time than would be needed for a full-scale implementation. Substantially fewer resources are at risk during a test. If the strategy turns out to be flawed for some reason, there should still be sufficient time and resources available to move in a different direction.

Testing candidate strategies usually involves a small-scale or trial implementation. For example, a new product might be introduced in a small segment of the company's potential market, or a new way of doing business could be implemented in one part of the organization. Some testing can be done with analysis. For example, a strategy that includes a major modification to the production process could be tested using a computer simulation of the production system. Consultants might be needed to help develop the simulation, but it would allow the costs and benefits of the change to be realistically evaluated. The simulation would also identify potential design or implementation problems that would need to be resolved during detailed planning for the change. (Simulation modeling is discussed in chapter 3.)

Testing can be the basis for an incremental approach to strategy implementation. This approach can significantly improve the chances of success when the firm is pursuing a strategy that requires broad changes in the way the business operates. It is very difficult to foresee all of the implementation problems that might arise when major changes are being made. A testing approach allows the organization to learn how best to implement the strategy. Implementation problems can be identified in the test and corrected before proceeding to full-scale implementation. This will reduce the time, cost, and disruption associated with the final implementation.

To be most useful, the strategy test must be designed carefully. If the strategy involves primarily internal changes, the test can be set up using the learning approaches to technology implementation described in chapter 6. The same principles can be applied to a strategy that is externally focused, such as introducing a new product or entering a new market. Key factors that need to be considered when designing the test include

- What are the key performance indicators for the test? What outcome do we expect? What are success/failure criteria?

- How representative is the test situation of the total domain over which the strategy will ultimately be implemented? How does the test "scale up?" For example, is it better to test market across a broad range of customers in a narrow geographic area or across a narrow range of customers in a broad geographic area?

- What information needs to be collected to ensure that the results of the test can be understood? How will this information be collected?

- What resources (financial, staff, facilities, and so on) will be required to conduct the test? Can these resources be made available without creating unacceptable impacts on current operations?

- How long will it take to set up and conduct the test? Will results be available in time to guide decisions about full implementation?

- Will the test impact existing customers? Can these impacts be controlled to acceptable levels?

- What are the risks that the test will reveal important strategic information to competitors?

- Are there other learning opportunities available from the test? For example, could more be learned about why customers buy current products at the same time that a new product is test marketed?

The strategy team can use these sorts of questions to identify several possible ways to test the strategy and select the test that provides the best information while keeping costs and risks at an acceptable level.

Testing is an approach more than it is an event. Sometimes more than one test or incremental implementation step will be needed. If the first test identifies a number of tricky implementation problems or potential difficulties with the strategy itself, it may be necessary to develop and test a modified strategy or implementation approach before moving ahead. If the changes being made are particularly complex, a series of implementation steps may be used, with tests at the beginning of each major step. If the strategy involves making significant changes in the interface between the company and its major customers or suppliers, joint tests should be conducted with the affected external organizations. Joint tests will ensure that systems in both organizations mesh after the change is completed.

Implementing the Strategy

After sufficient testing has been carried out to confirm the strategy and identify the most effective implementation approach, full-scale implementation can begin. Implementing the strategy can involve a variety of steps, depending on the nature of the strategic move. It could involve activities such as designing, purchasing, and installing new equipment; hiring new staff or training current staff; developing new products; changing company policies and procedures; developing relationships with new customers; or strengthening relationships with an existing customer by establishing a strategic partnership. These activities could move at a rapid pace or follow a more evolutionary approach. Regardless of the details, two things are usually important to initiate a successful strategy implementation: (1) develop an implementation plan and (2) communicate the strategy and implementation plan throughout the organization.

The implementation plan will outline the implementation steps, assign responsibilities, commit resources, and lay out the mechanisms that will be used to track the performance of the strategy. The implementation steps outlined as part of the evaluations performed to select the strategy provide the starting point for the implementation plan. The knowledge gained when the strategy was tested provides the information needed to fill in the details of the plan. The purpose of an implementation plan is not to develop elaborate documentation, but to make sure the strategy team thinks through the implementation process. In addition to being a good tool to ensure appropriate attention has been paid to the details of implementation, the plan can also be a vehicle for obtaining the commitments required to successfully implement the strategy. A sign-off process on the plan can be used as a device to formalize the deal among the managers and staff who will need to commit time, money, and staff resources to implement the plan.

The plan itself can serve as a good communications tool to help staff across the organization understand the strategy and how it will be implemented. Other communications mechanisms are also needed, such as informal presentations by managers, question-and-answer sessions, and articles in the company newsletter. Although some caution must be exercised to guard against divulging information that could be used by competitors, it is important to communicate the essence of the strategy broadly across the company if it is to be implemented successfully. People cannot be expected to implement effectively what they do not understand.

A key principal of the learning approach to strategy implementation is to move only as far as you need to, preserving future options when possible. Many of the best strategies will evolve through time through a process of testing, refining, and making incremental strategic moves. This does not mean managers should be risk-averse or wishy-washy. It does mean that major actions are not taken until they need to be taken and enough information is available to provide a high level of confidence that it is the right move to make. The key is to stay the course unless there is a good reason to change, but to move decisively when change is needed.

As mentioned previously, setting strategy is not an event, but a continuous process. It involves continuously collecting and synthesizing strategically important information and monitoring the performance of the current strategies. It requires that strategies be refined over time to keep them fine-tuned to events as they unfold. If conditions change significantly from those used to develop the strategy, a complete reassessment is warranted. The need for a comprehensive reexamination of the strategy can be identified by monitoring some selected strategic indicators. Most strategies are built on a few key things that are expected to occur in the future. Once a strategic course has been set, the strategist can monitor these key factors to see if they are, in fact, coming about. A scenario-based approach is particularly conducive to the identification of strategic indicators. These indicators are usually readily apparent from an examination of the scenarios and the potential outcomes. It is particularly important to monitor indicators that can give advance warning that either a particularly dangerous scenario or a particularly favorable scenario is beginning to unfold. These indicators would trigger an immediate reassessment of the current strategic direction.

Strategic Partnering

There are both advantages and disadvantages associated with being a small to medium-sized business. The advantages associated with smaller size include the ability to respond quickly as a single unit, simpler management and administrative systems resulting in lower overheads, and closer ties between management and staff. Disadvantages include the limited ability to absorb temporary financial setbacks; lack of specialists to advise managers on key areas of strategy; and difficulty getting full information about market forces, economic trends, and other strategic factors important to the future of the firm. Strategic partnering is an effective way to mitigate these disadvantages.

A strategic partnership is a long-term, mutually beneficial relationship between two organizations. It represents a commitment by the two organizations to work toward common objectives that can be achieved more successfully together than independently. Each firm involved in the partnership must give up a small amount of autonomy; but, by forging strong linkages with organizations that are key to its future, each firm significantly increases its chances of being successful over the long haul.

The most common type of strategic partnership for a small to medium-sized manufacturing business is a partnership with a major customer. This relationship provides the small firm with assurance of a long-term market for its product, access to a broader range of markets than it could access working independently, a chance to grow with the larger firm, and an opportunity to learn from a more experienced organization. The large firm gets an assured source of supply, rapid response when its needs change, and the confidence that its key supplier will move with it through strategic changes.

The strategic partnership may evolve over time as two organizations grow more and more interdependent. It can also be initiated as a proposal from one prospective partner to the other. In this case, one of the potential partners would come to the conclusion that there are significant mutual benefits available to the partnership and try to demonstrate this to the other partner. In either case, there will be a point in time when both partners recognize the benefits of a strategic relationship and declare their intentions to become partners. At this point, there will usually be some kind of formal declaration of the partnership. The declaration is generally not as formal or binding as a contract, but might be written down in a letter of agreement that describes the expectations and responsibilities of the partners. It may also lay out ground rules under which the partnership will be conducted and could spell out some of the mechanisms that will be used by the partners to define and achieve their mutual goals.

The partnership will operate through a number of strategic linkages. Through these linkages, the partners tie together elements of their key business processes. These linkages will generally exist across a broad range of activities from strategic planning, marketing, product development, and production processes to finance and accounting systems. Through these linkages, the partners are able to operate almost as if they were a single organization. The linkages permit the partners to enhance their agility because they can move together to respond more rapidly to changing conditions in the marketplace. These linkages also enable the partners to streamline shared business practices and work processes, thereby lowering costs. Linking mechanisms that might be used include the following.

• *Establish close working relationships between counterparts in key positions in each firm.* This may be a preliminary to the formalization of the strategic partnership, as part of the process to confirm that the partnership makes sense for both parties. These personal relationships will be especially important in the early stages of the partnership, as the partners are learning more about each other.

• *Assign an on-site representative to work full time at a partner's key facility.* Much of the coordination required to make the partnership work well can be done electronically, but if the partners are located some distance apart, the on-site representative can be a useful mechanism to keep things working smoothly. This is particularly important if the partners are several time zones apart or have very different work schedules. The on-site representative is always there to deal with coordination issues. On-site representatives probably should be rotated every six to 12 months to keep them current in their home organization.

• *Link information systems.* Linking electronically enhances communication by making key information available whenever and wherever it is needed. Many routine transactions between the firms can be automated, saving time and money. Computer-integrated manufacturing and similar methods can be adopted, dramatically improving the response time and lowering the cost of the product life cycle. Electronic linkage also enables shared production planning and control systems to be put in place. Each partner is able to "see" the capacity planning and master scheduling information for the other's production processes, permitting the partners to integrate their production systems. The integrated systems should result in lower production cycle times, improved on-time delivery performance, and reduced production costs.

• *Share information on key performance metrics.* Such sharing will provide early indications of actions the partners need to take together, or indicate if an independent action of one partner could indirectly impact the other. The partnership should also track joint performance measures that show progress toward the goals they establish together.

• *Conduct joint product development activities.* These joint activities are particularly important if the smaller partner provides components for products the larger partner sells in the marketplace. Joint product development will help the team bring the new product to market faster and will also enable producibility, capacity, and other considerations associated with the smaller partner's production systems to be factored into product development decisions.

As the relationship evolves, the major customer will likely become a key source of strategic information for its smaller partner. Because of the strong interdependencies that develop within the partnership, the larger firm has strong incentives to share information about market trends, economic trends, and technology developments applicable to both the products and the production processes of the smaller firm. The larger partner may even share the investment costs for new equipment that will result in lower costs, quality improvements, or more reliable delivery schedules for the products it uses. Shared strategy setting is a key factor in the long-term success of the partnership. Unless the partners share key strategic goals and implementing strategies, they cannot have a full partnership and achieve all of the potential benefits. In a mature partnership, each partner will participate on the strategy-setting team of the other partner.

There are risks associated with strategic partnerships. These risks should be recognized and managed as the partnership is formed and maintained. The major risk is that the smaller firm will be impacted by any problems its larger partner experiences. For example, if a strike or natural disaster shuts down the portion of the partner's production process that uses the smaller firm's product, the smaller firm will immediately experience revenue losses. If the larger partner has problems in another part of its business, it could reduce the resources available to pursue the business agenda of the strategic partnership, resulting in slower growth for the smaller partner. These risks exist to some degree for any small to medium-sized company that sells a large fraction of its output to a single customer. The smaller firm will be impacted by these kinds of events, even if it is not in a strategic partnership. It is, in fact, because of these risks that the smaller firm has incentives to pursue a partner relationship. If the two firms have formally recognized their strategic alliance, the larger firm will be more likely to try to mitigate the impacts of these events on its smaller partner. The larger firm will recognize that the impact could be severe on the smaller partner and look for ways to reduce those impacts and preserve the longer-term benefits expected from the relationship.

Another potential risk associated with the strategic partnership is that the larger firm will dominate the smaller firm. If this happens, the smaller firm becomes a de facto subsidiary of the larger firm. Because the smaller firm is so strongly dependent on the revenues from the larger firm, it may feel it has no way out of this relationship. Again, these risks are present in any business situation in which a smaller firm is strongly dependent on a single customer, with or without a strategic partnership. There is no magic solution for this problem. The best approach is to avoid the situation by choosing partners carefully. If the potential partner is not an organization that can be trusted, aim the business in a different direction.

Although strategic partnerships with a key customer are the most common, it is also possible to develop a partnership with a key supplier. A company generally works with several categories of suppliers—those who provide materials and services that go directly into its product, those who provide services needed to operate the business, and those who provide production-related equipment. Suppliers who provide product-related materials and services will generally be the best candidates for strategic partnering arrangements. The best candidates will usually be larger firms, as partnerships with small suppliers won't help mitigate the disadvantages associated with being a small to medium-sized firm. A strong relationship with a supplier of production equipment also can be very beneficial. Through this relationship, the small firm can work with the supplier to develop equipment that is tailored to meet its needs. This can be done even when the small firm is not one of the larger customers for this supplier.

If the smaller firm is representative of companies that form a significant part of the supplier's business, a partnership can help the supplier do a better job of meeting the needs of many of its customers. The small firm brings a realistic test bed to the partnership and receives the benefits of an early start in applying improved production technology to its key processes. Many of the linking mechanisms described for establishing and maintaining partnerships with customers also can be used with key suppliers.

A third type of strategic partnership is to develop a long-term relationship with an R&D organization. Many large companies are finding it is not cost-effective to maintain their own R&D staff. They are beginning to outsource much of their R&D to companies that do R&D as their only business. This approach is even more applicable to a small to medium-sized company. It will work best to establish a long-term relationship so the R&D partner gets to know your business. The R&D partner will not only be able to respond with the right expertise when requested, but will also proactively bring new developments to the attention of the firm. A good R&D partner can also keep its clients up to speed on relevant technology developments for a modest cost through technology assessments or group projects that provide information to a group of firms with similar interests.

Conclusion

The agile firm will utilize strategic practices that enable it to be successful in spite of the uncertainties surrounding future events. The firm will have the strategic agility it needs to a be successful across a broad range of future market conditions. The foundation of strategic agility is participation in strategic processes by all members of the organization. The managers lead, but everyone has a part in collecting the best available strategic information, finding the best strategies, and implementing them effectively and efficiently. A learning approach to strategy enables the firm to find the proper balance between uncertainty and the need for action, between analysis-based and intuition-based strategic decision making, and between evolutionary changes and decisive strategic moves. Strategies are implemented by the agile firm by aligning all the parts of the organization so it moves as a unit to achieve the strategic goals. The agile small to medium-sized manufacturing firm will make effective use of strategic partnerships to mitigate the disadvantages associated with its size. Strategic partners will share the load and significantly increase the chances of success.

References

Drucker, P. F. 1994. The theory of the business. *Harvard Business Review,* 72 (5): 95–104.

Hamel, G., and C. K. Prahalad. 1989. Strategic intent. *Harvard Business Review* 67 (3): 63–76.

Schwartz, P. 1991. *The art of the long view.* New York: Doubleday.

Additional Reading

Hax, A. C., and N. S. Majlaf. 1991. *The strategy concept and process: A pragmatic approach.* Englewood Cliffs, N.J.: Prentice Hall.

Montgomery, C. A., and M. E. Porter, eds. 1991. *Strategy: Seeking and securing competitive advantage.* Boston: Harvard Business Review Book Series.

CHAPTER 8

Performance Measures

Joseph C. Montgomery and David J. Lemak

Measuring organizational performance is a key step toward achieving organizational effectiveness (Hall, Johnson, and Turney 1991). Without performance measurement information, the organization has no way of knowing whether it is achieving its mission, whether its strategies are appropriate to the situation, or whether implementation of the strategies is being successfully accomplished. Given the critical importance of performance measures, this chapter attempts to (1) make the case that performance measures tend to be based on a ineffective business paradigm that does not provide the information needed for world-class performance, (2) show how such measures drive short-sighted behaviors and ineffective performance, and (3) develop a model for performance measures (based on internal and external customer feedback) that will encourage world-class performance.

The role of performance measures in the context of organizational mission, vision, strategies, and actions is indicated in Figure 8.1. According to the figure, the mission of the organization—the definition of the organization's purpose, business areas, and value-added activities—creates the need for a clear management vision of how best to achieve this mission. The vision drives the development of organizational strategies and plans. These strategies and plans, in turn, guide and determine the actions to be taken by organizational members.

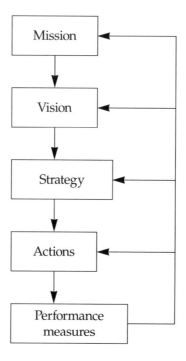

Figure 8.1. Role of performance measures in organizational effectiveness.

Actions include creating and delivering products and services, establishing and maintaining effective work processes, designing the organizational structure, formulating policies, and so forth. These actions direct the development of appropriate performance measures to serve as indicators of organizational performance. Performance measures therefore provide a framework for the systematic collection and analysis of data summarizing performance within the organization.

A variety of measures have been used by manufacturing organizations. Historically, measures have included return on investment, market share, efficiency, labor utilization, and machine utilization. More recently measures have included cycle time, waste rate, setup time, customer satisfaction, response time, quality, output, and so forth. These clearly different types of measures reflect radically different underlying approaches to operating a factory. The next section outlines the origins of the first set of measures.

A Historical View of Performance Measurement

Performance measures have traditionally been developed from a cost-accounting orientation, also called a manager-centered approach (Hall, Johnson, and Turney 1991). Cost accounting was developed in the late 1800s to help managers evaluate the total costs of operating textile mills, railroads, steel mills, retail stores, and similar businesses (Kaplan and Johnson 1987). Cost was divided into two basic categories: (1) variable costs, such as labor and materials, that might vary with market rates, and (2) fixed costs, including maintenance and facility upkeep, that were expected to be fairly stable over time. At that time, variable costs formed a considerable portion of total costs, perhaps 80 percent or more, as salaries and materials formed the greatest expense. Cost accounting basically involved breaking the production line into its departments or functions and allocating charges to each department or function to cover the associated variable costs. As long as this was done on a roughly rational basis, the result was a sensible way of tracking costs. Cost-related measures were also developed for each department/function to help assess how company funds were being spent. Standard financial indicators resulting from the cost-accounting approach included machinery utilization, labor utilization, unit cost, and inventory turnover. Overall organization-wide cost-accounting measures included net profit and return on investment.

Unfortunately for this financially oriented approach, production methods and technologies have changed dramatically, particularly in the past two decades, and the underlying assumptions have become questionable. For example, variable costs have dropped from 80 percent of the total until now they account for only 10 percent to 15 percent of total costs. Labor costs, for instance, have become only a small fraction of total operating costs in a factory. As a result, relying on these financial measures means managers are paying attention to what have become relatively minor issues. What was formerly an appropriate methodology has evolved to a case of "the tail wagging the dog."

As will be demonstrated in the next section, financial measures motivate patterns of behavior that interfere with efficient and effective production processes. Cost accounting–based performance measures make it very difficult for organizational members to keep the overall mission in mind and to avoid succumbing to strong pressures to behave in ways that hurt overall organizational performance.

Dysfunctional Impacts of Cost-Accounting Performance Measures

The old saying, "you get what you measure" has proven to be the case, perhaps to a far greater extent than is generally appreciated. Performance measures, once created, tend to take on a life of their own, serving as powerful motivators of behaviors within the organization (Dixon, Nanni, and Vollman 1990; Kerr 1995; Scott-Morgan 1994). When cost accounting–based performance measures are used, the relationships shown in Figure 8.1 quickly devolve to those shown in Figure 8.2. In this situation, the organization mission and vision become disconnected from strategies, actions, and performance measures. The performance measures—or perhaps more correctly, the drive to perform well on the measures—drives both strategies and actions. The cost-accounting measures begin to drive both organizational actions and strategies because they (1) provide information to organization members as to what is important and (2) indicate to groups and individuals how their performance will be evaluated. Consequently, these performance measures become extremely powerful motivators of behavior throughout the organization (Kaplan and Johnson 1987). Unfortunately, in the short-sighted drive to obtain good results on the performance indicators, organization members tend to lose sight of the organizational mission and vision. The mission and vision, therefore, are disconnected from organizational actions and no longer serve as an orienting and guiding force.

Once cost-accounting measures are in place, organizational members in each department are motivated to "look good" on these measures. In order to do so, they work to maximize output of the unit, minimize unit cost figures, and maximize performance on other measures, such as machine and labor utilization. These efforts are pursued in isolation—the performance of other units is not a concern because only individual unit performance is reflected in the measures. Because these actions are pursued without consideration for the next step in the

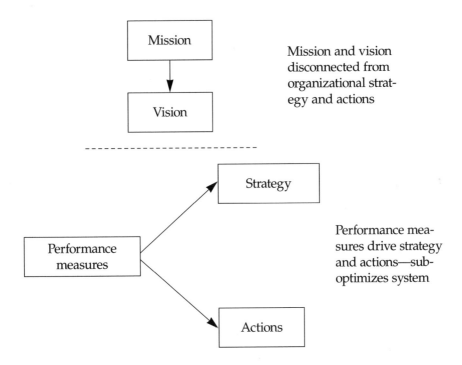

Figure 8.2. Effects of cost-accounting measures on organizational effectiveness.

work process, a lack of alignment of activities, poor communications, misunderstandings, and outright competition begin to occur. Research has, in fact, shown that whenever subunits work to achieve their individual best performance, performance of the organization as a whole will suffer (Senge 1990).

In addition, many of the very actions that will make a unit look good on its performance measures will directly harm production. Let us consider a concrete example using a single production line. The management of the line is aware that its performance is being monitored using such measures as machinery utilization and direct labor utilization. Therefore, the more equipment and workers are kept busy, the more effective the unit appears. Relying on long production is a great strategy to help keep these measures high. Long production runs also seem to be a logical way to maintain productivity.

Unfortunately, the long production run strategy has a number of unanticipated side effects (Schonberger 1990). For example, an immediate result is to create significant amounts of work-in-process (WIP) to clog the line. At the end of the line, excess inventory begins to pile up as loading and shipment fail to keep pace with the high level of output. In the drive to sustain production, quality problems become difficult to detect, diagnose, and correct. Overall, quality begins to deteriorate.

The long production runs also mean there is little time available for equipment maintenance and for the development and training of personnel. Performing maintenance and setting aside time for training also cause a negative impact on other measures, such as direct labor utilization. Thus, management provides only reluctant approval for preventive maintenance and training. Numerous problems result. Breakdowns become more common as equipment is pushed beyond intended limits. Workers begin to fall behind technically and become less capable of performing their jobs because training has been decreased.

Given these problems, it is only natural that resources devoted to rework increase and scrap rates rise. Storage, transportation within the production area, and work scheduling become increasingly difficult. The long production runs also mean that products cannot be tailored to meet the needs of individual clients or rapidly changing markets. Therefore, the products become increasingly less desirable to customers and the competitive situation of the company begins to deteriorate. The overall result is chaos, waste, incredible inefficiency, and loss of strategic competitive position—all created by the well-intentioned but short-sighted efforts by management and staff to demonstrate good performance on the cost-accounting measures.

At the departmental or unit level, the decisions dictated by cost-accounting measures appear to be rational ones. However, at the organizational level, these decisions are devastating. The production characteristics that emerge when cost-accounting measures drive production—large batch sizes, long production runs, limited numbers of products, slow changeovers, high overhead costs, poor quality, and long lead times—are *exactly the opposite* of those recommended by world-class manufacturing experts (for example, Black 1991; Schonberger

1990, 1992; Suzaki 1987). How ironic that focusing on costs ultimately results in higher costs and less-satisfied customers!

The Customer-Centered Paradigm

Performance measures used by world-class organizations tend to be customer-centered, rather than based on the older cost-accounting approach (Hall, Johnson, and Turney 1991). Customer-centered performance measures are linked with product quality, dependability of service, waste reduction, timeliness, flexibility, innovation, and other indicators that are closely linked with actual work processes. Development and implementation of these measures has often resulted in dramatic improvements in internal work effectiveness and in the performance of products and services in the competitive marketplace (Young 1992).

The difference in orientation and actions of organizations that have a customer versus a cost-accounting orientation is so great that they have been labeled as separate operational paradigms. A variety of familiar programs fall into the customer-centered paradigm, including TQM, JIT manufacturing, total productive maintenance (TPM), continuous improvement, CIM, total employee involvement, and quality function deployment.

Key differences in assumptions between the cost-accounting and customer-centered paradigms are summarized in Figure 8.3. As can be seen, the cost-accounting paradigm describes the traditional U. S. organization.

Numerous writers (for example, Suzaki 1987; Schonberger 1990) have presented detailed arguments as to why customer-centered organizations outperform those with a cost-accounting orientation. Those focused on cost-accounting and associated assumptions simply cannot compete with customer-oriented organizations in such areas as innovation, quality, timeliness, cost, and customer satisfaction.

In contrast, a much different picture emerges when customer-oriented performance measures are used. For one thing, Figure 8.1 continues to hold true—the mission and vision remain connected to and drive strategies and actions. In addition, customer-oriented measures build upon the concept of an internal chain of customers within

Cost accounting
- A company is its assets; the company is a possession.
- Economies of scale—bigger is better.
- Managers manage, workers work.
- Management initiates improvement.
- Vertical, top-down organization; functional silos separate groups.
- Profit is first, cost/trade-off thinking prevails.
- Company-centered operations.
- Transaction-driven management.
- Performance measurement for control; financial measures dominate.

Customer-oriented organizations
- A company is its people.
- Economies of scope—faster response to customer needs is better.
- Workers are thinkers; everyone works for improvement.
- Horizontal organization, multidirectional communications.
- Internal chain of customers, strong links with external customers and stakeholders.
- Quality is first priority. No compromise with quality and customer service.
- Manufacturing-centered operations.
- Improvement-driven, problem-solving teamwork orientation.
- Customer-oriented performance measures are dominant. Performance measures guide work improvement and plant operations.

Figure 8.3. Assumptions of cost-accounting versus customer-centered organizations.

the organization (Schonberger 1990) and help to maximize overall organizational performance. For example, if a production process involves a series of steps by different work groups, then a series of supplier–customer relationships exists that links the process from beginning to end. These relationships are shown in Figure 8.4. Performance measurement data collected from the first customer group by the initial supplier group might include timeliness, quality, defect rate, and perceived customer-service orientation of the supplier. The supplier group might also collect its own data on scrap rates, customer

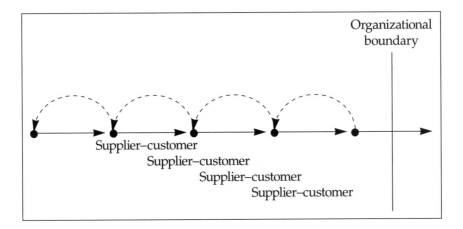

Figure 8.4. The internal chain of customers.

complaints, quality, and related variables. These data would be continually scrutinized, and indications of any problem would immediately lead to corrective actions. As a result, the initial supplier will provide the materials needed to its customer in the quantity needed and at the time they are needed. Unlike the case with cost-accounting measures, when a work group attempts to look good on its customer-oriented measures, the result is a win–win situation. Both the work group and the organization come out ahead.

As the initial customer becomes, in turn, the supplier for the second link in the chain, exactly the same process of performance measurement and feedback of data would be used. The use of customer-centered performance measures ultimately results in connecting each link in the production chain. Once these connections are made, the "horizontal organization" (Ostroff and Smith 1992) has been created, in which work products flow smoothly, efficiently, and effectively through the organization, unimpeded by functional boundaries.

Developing Customer-Based Performance Measures

The flow diagram shown in Figure 8.5 provides a practical methodology for developing customer-oriented measures. The starting point is

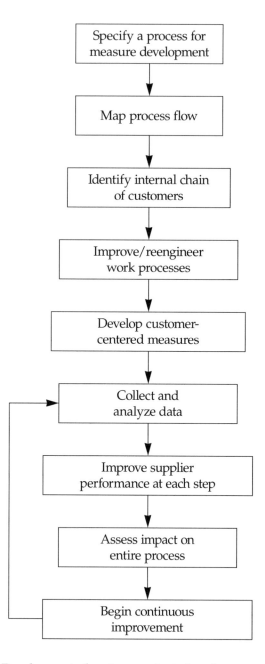

Figure 8.5. Development of customer-oriented performance measures.

selecting an initial process for which customer-oriented performance measures will be developed. This is not a trivial step, as a complete process runs from initiation to completion and typically crosses a number of organizational boundaries. Processes tend to be invisible and unnamed because people tend to think in terms of existing departments and structure, rather than the whole job (Hammer and Champy 1993, 118). Therefore, the team developing the measures must have a clear understanding up front about the definition of a process. Processes might include product development, sales, order fulfillment, customer service, or product assembly.

How should a process best be selected? Later steps in Figure 8.5 call for improving or even reengineering the process before developing the measures. Therefore, the process selected should be one that is important to the organization, has significant potential for improvement, and that—if improved—will make a real difference to the company.

The next step is to map out the process flow. This should involve, at most, a few weeks' time to identify and describe the major steps in completing the work that is done. The goal is a simple diagram that shows the main steps involved in the process with some accompanying text to describe generally what happens in each step and who is involved. The completed product should strike everyone as being obvious and an accurate reflection of the work. Because the diagram will cross organizational boundaries, this may represent the first time that the process has been examined from beginning to end. A word of caution—avoid getting involved in excessive detail. A high-level diagram is sufficient at this point.

After diagraming the process flow, both the internal and external customers, as well as key stakeholders, must be identified. The internal chain of customers may simply involve revisiting the process diagram and relabeling each box as a customer for the previous step and the supplier for the subsequent step. Organization members must understand that their customer may well be in the next office and that internal customer satisfaction and concern for quality are just as important as satisfaction of the external customer. Identifying the internal chain of customers sets the stage for developing relevant, valid performance measures. While this may seem to be a trivial step, it is not. This may well be the first time that the workers responsible for the functions or steps within the process have ever thought of themselves as either having customers or being one.

The next step in the model is labeled "Improve/reengineer work processes," which involves taking the time to correct glaring process problems or, if major improvements are required, to radically redesign how the work is performed. It makes no sense to develop and implement performance measures for flawed processes. (Refer to Hammer and Champy 1993 or Davenport 1993 for more information on reengineering. Note that changes in organizational structure may be needed as well to support the improvements.) Much of the rest of this book is, in fact, devoted to these sorts of improvements. Specific improvements may include reducing distance between workstations, reducing flow time and space along the customer chain, reducing setup and startup times, or creating a kanban system to allow the customers' rates of usage to pull materials through the system. Regardless of the techniques employed, it is critical that the improvements be initiated and driven by the organizational members and work teams most closely involved with the work processes.

After improving the process, performance measures are developed for each supplier–customer link. The third step identified suppliers and customers. This step requires the supplier staff to meet with customer staff to discuss customer requirements and brainstorm appropriate measures. The exact nature of the measures will depend on customer needs. Measures might center on quality, dependability, timeliness, and customer satisfaction. We recommend that members of the performance measures development team facilitate these meetings, for several reasons. First, it is likely that the two groups will never have met before to discuss how they work together. Some degree of mistrust or anxiety is likely. Second, the supplier–customer meetings may well involve serious negotiations about the measures and about the way the work is performed. Some aspects of the way the suppliers are currently performing are likely to create problems for the customers. Surfacing and discussing these issues may well be uncomfortable, particularly for the supplier staff. The facilitator's role is to allow an open discussion, but one without blaming or criticizing of other members. The goal is to come up with win–win solutions to problems and measures—solutions that are acceptable and beneficial to both groups. In general, each work team will be involved in at least two meetings, as they have the dual role of suppliers to the next step in the process and customers for the previous step.

Two groups will have somewhat different concerns: the initial and final steps in the process. Staff involved in the first process step will need to initiate measures with their own suppliers, who may include external vendors. Staff in the final step will need to initiate contact with an external customer, who may be within or outside of the organization. Other organizational groups, such as marketing or sales, may claim responsibility for external customer contacts. However, allowing the final process step team to contact customers directly can have significant payoffs in terms of customer satisfaction, worker morale, and empowerment of everyone involved in the process.

Once the measures are developed, data collection begins to establish baseline performance levels on these measures. The baseline data is analyzed by the supplier/customer teams throughout the process and plans are developed and implemented that will improve subsequent performance on the measures. Ideally, the data collection will be displayed openly to publicize and document process improvement (Hall, Johnson, and Turney 1991). It is critical that the measures be institutionalized into the organization, in the sense that performance appraisal, raises, promotions, and so forth be made contingent on successful performance on these measures.

The process owner (who may need to be named, if none currently exists) is responsible for assessing the overall impact of the improvements going on at each step in the process. That is, measures such as cycle time and waste may need to be "rolled up" across each step in the process to provide overall measures. A process for pulling together the data from each team will therefore need to be developed. Measures used in the final step in the process (for example, quality, total output) will also provide valuable information for the process owner.

As a result of these steps, organization members involved in the selected process will be well on the road to continuous process improvement. They will be continually gathering data to assess and improve performance for their customers and will be working with their suppliers to obtain similar improvements. Individual rewards, bonuses, promotions, and so forth are linked with these improvement efforts. As the new system settles into continuous improvement mode, organizational resources may then be focused on another process in need of measure development and improvement.

Conclusion

Customer-centered performance measures, unlike those based on cost accounting, result in improved organizational effectiveness in a variety of ways. These measures provide a powerful motivation to improve quality, reduce cycle time, develop innovations, and generally improve the value-added activities of each supplier. When performance measures are developed that link each supplier and customer in the customer chain, the organization effectively becomes horizontally oriented. Traditional functional silos and organizational barriers are eliminated.

Customer-centered measures provide the information that management needs to make operational decisions. Experts in the area of production and productivity strongly advise managers not to make operational decisions based on cost-accounting data (Dixon, Nanni, and Vollman 1990; Hall, Johnson, and Turney 1991; Kaplan and Johnson 1987; Peters 1988). This is not to say that there is no role for cost-accounting measures or for measures associated with other approaches, such as activity-based accounting. Cost accounting measures may well be required for external reporting purposes, perhaps to regulatory agencies or to fulfill legal requirements. Activity-based accounting, a more recent derivative of cost accounting, may be very useful for strategic decision making, such as whether to pursue production of a certain product or service. We argue strongly against using either cost-accounting or activity-based–accounting measures to guide the actual operation of an organization. The cost in terms of dysfunctional behaviors and loss of mission orientation is simply too great.

It is admittedly a temptation for managers to desire a small set of financially oriented measures upon which to decide whether their organization is functioning effectively. Such measures, along with quick-fix solutions to difficult problems, would certainly make the management of complex organizations far easier (Kilman 1991). All that can be offered is assurance that once customer-centered measures are in place and continuous improvement efforts are increasing the effectiveness of work processes, the values of certain key cost-accounting measures will gradually reflect the improvements.

Some financial measures may at first display alarming trends. For example, increased inspection of purchased items may be needed early on, so overhead costs for the purchasing organization may rise. Machine utilization and direct labor utilization may decrease, overhead needed for training may increase. However, as the process becomes more effective, and as other processes are included in the program, overall financial performance and profitability should improve dramatically. Top management will need to display patience until these improvements occur and may need to protect changes generated by the customer-centered approach from outside interference. (Refer to chapter 2 on change management for assistance in dealing with such issues as resistance to change.)

Finally, we want to emphasize that many financial measures, including direct labor utilization and machine utilization must be seen as *permanently irrelevant* to organizational functioning. In modern manufacturing, worker downtime may simply reflect a trouble-free workflow and will provide time for innovation and preventive maintenance. Similarly, excess machine utilization suggests overuse and inadequate maintenance. Many other financial measures will also be found to have no merit whatsoever in the new way of operating. In any event, whether cost accounting–based measures are irrelevant or of some use, managers must resist the temptation to use these measures to guide operation of the organization.

References

Black, J. T. 1991. *The design of the factory with a future.* New York: McGraw-Hill.

Davenport, T. H. 1993. *Process innovation.* Boston, Mass.: Harvard Business School Press.

Dixon, J. R., A. J. Nanni, and T. E. Vollman. 1990. *The new performance challenge: Measuring operations for world-class competition.* Homewood, Ill.: Richard Irwin.

Hall, R. W., H. T. Johnson, and P. B. Turney. 1991. *Measuring up: Charting pathways to manufacturing excellence.* Homewood, Ill.: Richard Irwin.

Hammer, M., and J. Champy. 1993. *Reengineering the corporation.* New York: Harper Collins.

Kaplan, R. S., and H. T. Johnson. 1987. *Relevance lost: The rise and fall of cost accounting.* Boston, Mass.: Harvard Business School Press.

Kerr, S. 1995. On the folly of rewarding A, while hoping for B. *Academy of Management Executive* 9 (1): 7–16.

Kilman, R. H. 1991. *Managing beyond the quick fix.* San Francisco: Jossey-Bass.

Ostroff, F., and D. Smith. 1992. The horizontal organization. *The McKinsey Quarterly* 1: 148–68.

Peters, T. 1988. *Thriving on chaos.* New York: Alfred Knopf.

Schonberger, R. J. 1990. *Building a chain of customers.* New York: Free Press.

———. 1992. Is strategy strategic? Impact of total quality management on strategy. *Academy of Management Executive* 6 (3): 80–87.

Scott-Morgan, P. 1994. *The unwritten rules of the game.* New York: McGraw-Hill.

Senge, P. M. 1990. *The fifth discipline.* New York: Doubleday.

Suzaki, K. 1987. *The new manufacturing challenge: Techniques for continuous improvement.* New York: Free Press.

Young, S. M. 1992. A framework for successful adoption and performance of Japanese manufacturing practices in the United States. *Academy of management review* 17 (4): 677–700.

CHAPTER 9

Creating the Learning Factory
Linda R. Pond

Magazines, journals, and newspapers are filled with articles pointing out the need for businesses, from giant corporations to small manufacturers, to operate in a constant state of learning. When authors attempt to describe the concept of continuous learning they use many different terms for the idea of a business that incorporates learning as a regular part of its operations. For instance, what will be referred to in this chapter as a *learning factory* has been described by other authors as *learning businesses, knowledge organizations, centers for learning, total quality learning organizations, companies built on knowledge workers,* and so forth. In spite of the different labels, a common theme in learning-focused discussions is a warning that those companies that do not become learning organizations will soon face corporate death. This sends a powerful message and is a troublesome idea for many to believe or accept.

These warning messages may gain credibility by considering the case Davis and Botkin (1994) make for becoming a learning organization. They see the next wave of economic growth as coming from knowledge-based business, and, similar to Peter Drucker (1993, 3–18), they observe that business becomes knowledge-based by putting information to productive use. They clearly express the impact of knowledge as they remind readers that in agricultural-based economies, learning is often promoted by religious groups, focuses on children between the ages of seven and 14, and is sufficient to last all the years of a working life. In industrial-based economies, learning is traditionally led by government, reaches students between the ages of five and 22, and, again, is sufficient for most working lives. On the other hand, in knowledge-based economies, they assert that learning must be constant and education must be updated throughout one's working life. People must continue developing their knowledge base—they must continue learning in order to increase their productivity and earning power.

Davis and Botkin pose the sobering thought that information (what people need to know) is doubling about every seven years, and in technical fields about half of what students learn in their first year of college is obsolete by graduation. They predict that over the next few decades, the private sector will take over the role of education that is currently government sponsored and within the public sector. Even today some public school districts, such as the Baltimore, Maryland, school district, are contracting with private firms for management of their school system, including management of the school district, curriculum development, and education methodology. Similar contracts have been let for public school educational services in Miami, Florida, and other areas of the country. Expanding on the idea of privatizing public school education, Davis and Botkin envision that businesses, including manufacturers, will become the educators both of their own employees and of their customers. Stepping up to the challenge of educating employees and customers is an important part of being a learning organization. It enables the manufacturer to set the direction for the future and use a skilled, knowledgeable workforce to focus on the educated customer's requirements. Education and learning provide a forum for anticipating and understanding the customer's needs, and can facilitate agile responses to changing needs.

For the manufacturers whose businesses are thriving, who have built strong training programs, hired good people, purchased quality equipment, and developed outstanding customer relationships, the idea of pending loss of business and, ultimately, failure because they are not also a learning organization may seem far-fetched. These manufacturers have heeded strong industry messages to study what successful Japanese manufacturers have done. They have implemented TQM strategies, including using statistical process control (SPC) methods and quality circles to ensure that problems are quickly identified and solved. They have set up kanban systems in their factories and have reduced cycle time. They are using JIT methods as a regular business practice and may be in the midst of reengineering their business and production processes. They have increased the amount and kinds of training offered to employees, and are constantly on the lookout for better training methods. Now manufacturers are being told that they also must be learning organizations in order to thrive or, indeed, survive. Understanding how continuous learning is part of the business strategy for successful factories is essential to their success.

Being a Learning Factory Is an Imperative for Success

Understanding the benefits of a learning organization means sorting out where or how the idea of learning fits with other survival strategies such as benchmarking, downsizing, streamlining processes, and concurrent engineering. Some of the earlier literature on learning organizations, for instance Peter Senge's *Fifth Discipline* (1990) or Argyris and Schon's *Organizational Learning* (1987), provides significant concepts and theories about learning organizations, but does not provide a clear pathway for implementing the concepts.

Senge et al. (1994) talk about companies building learning organizations because they want to achieve superior performance, improve their quality, satisfy customers, gain competitive advantage, energize and develop a committed workforce, and so forth. It's hard to think of a manufacturer who does not want to do these things and it is not clear how becoming a learning factory will achieve those desires. The authors make a stronger point for becoming learning factories when they state, "In the long run, the only sustainable source of competitive advantage is your organization's ability to learn faster than its competition. No outside force can take the momentum of that advantage away from you." (Senge et al. 1994, 11). Peter Drucker (1993, 3–18) adds that in our society knowledge is the primary resource for individuals and for the economy. He says that knowledge becomes productive only when integrated into a task; in other words, the value from knowledge is gained when the knowledge is used. The factory owner who has a learning organization has the opportunity to increase productivity through the use of knowledge.

Others who echo the belief that manufacturers must become learning organizations say that knowing what it takes to be the best today will be inadequate tomorrow. Bowen et al. (1994) share this belief and report that their studies indicate that in the manufacturing world of the 1990s, the key to success will be to excel in learning and convert that learning into commercial products and processes.

How Does a Factory Become a Learning Factory?

In order to discuss how to become a learning organization, we need to have a shared understanding of what we mean by *learning organization*. Garvin (1994) provides a clear definition that can be used for our shared understanding: A learning organization is an organization skilled at creating, acquiring, and transferring knowledge, and at modifying its behavior to reflect new knowledge and insights.

For manufacturers to remain competitive, even in the next few years, they will have to become learning organizations. For many this implies changing focus from equipment, processes, and technology to also include people who are continuously learning and applying knowledge. It is important to understand that this means shifting from considering training to be the primary means for gaining knowledge and skills to becoming an organization whose processes and systems take advantage of learning opportunities at every possible level. Workers must apply their newly gained knowledge to implement process and product improvements, and to use technology and equipment more productively. In a learning environment, workers look ahead to what else is needed—things such as, how is information communicated, how is knowledge gained and applied, what are better ways of doing business, what do consumers want, and how will those needs be met.

One example of looking ahead is a discount food warehouse that changed its checkout system from one where checkout clerks had almost no opportunity to talk with or listen to the customers to one where the actual checkout process is highly automated, thus giving the clerks time to talk with customers and learn what their needs are as well as how they can be served better. The clerks can function as a "look ahead" resource as they learn what customers expect and convey this to management. The clerks are becoming knowledge workers. They are expected to perform more than a checkout function—they are part of the effort to communicate with the customers, learn their preferences, anticipate their needs, and process the information in a way that can be used for improving service.

An important issue to consider when deciding to transition from a traditional factory to a learning factory is that there is no clear, proven road map for how to make the transition. Businesses searching for models to emulate or companies to benchmark because they are noted for their success at becoming learning organizations will have difficulty finding appropriate models or benchmark targets. Some companies, such as Xerox, are making this transition, but realistically, becoming a learning organization is plotting new territory.

We've evolved from being an agricultural society to an industrial society to a technological society. Now we are becoming a consumer society. The characteristics of a consumer society require that manufacturers anticipate and rapidly respond to market demands. Everyone is familiar with the concept of "faster, cheaper, better." A key to that concept is being a learning organization; unfortunately, the textbook for doing that hasn't been written yet. In spite of the lack of a textbook or road map to help facilitate the transition from traditional to learning-based manufacturing, there are elements of a learning organization that are considered by many researchers to be critical, and these can help manufacturers develop their own road map for making the transition to a learning organization.

Building a Road Map for Becoming a Learning Factory

Bowen et al. (1994) provide a template that includes elements considered by many researchers to be essential for becoming a learning organization. As you review these elements, consider where your factory is in relation to them. Assess where you need to be, and how you can incorporate the elements into your own organization.

Core Capabilities

Core capabilities are those special capabilities that allow a manufacturer to serve customers in a way that distinguishes the manufacturer from other manufacturers. Core capabilities are a mix of knowledge, skills, systems, processes, and values (the manufacturer's predominant attitudes, norms, and behaviors) that define an organization. Core capabilities allow a factory to compete based on matching its strengths to market requirements. As an example, Stalk, Evans, and Shulman (1993, 19–40) in *The Learning Imperative* report that Wachovia Corporation, a financial institution with headquarters in Atlanta, Georgia, and Winston–Salem, North Carolina, has outstanding returns and a growing market share in both Georgia and North Carolina. Whereas in Columbus, Ohio, Banc One, also a financial institution, has consistently earned the highest return on assets in U.S. banking industry. Both Wachovia and Banc One are highly successful, but both have achieved this success by leveraging off different capabilities.

Wachovia focuses on the individual customer. Its frontline employees, 600 "personal bankers," provide Wachovia's mass-market customers with personalized service that is usually associated with private banking clients. Wachovia uses specialized support services to enable the personal bankers to do their job. These systems include integrated customer information files, simplified work processes that allow the bank to respond to almost all customer requests by the end of the business day, and a five-year personal banker training program. Wachovia may have the highest cross-sell ratio (the average number of products per customer) of any bank in the country.

Banc One competes in a different way. The bank focuses on the entire community. To achieve this, it has developed strong community relationships and is perceived as part of the community. Banc One has a company slogan that says it "out-locals the national banks and out-nationals the local banks." At Banc One the focus is on the 51 affiliate banks in its network, rather than on personal bankers. Affiliate banks are led by presidents who have exceptional power within their regions. They select products, establish prices and marketing strategy, and make credit decisions as well as set internal management policies. This allows them to be an integral, responsive part of their community.

Manufacturers must stay vigilant to recognize when their core capabilities are transitioning from assets into habits. Bowen et al. (1994) warn against allowing core capabilities to turn into core rigidities. This danger is important to acknowledge and difficult to guard against. A manufacturer has to be keenly aware of what is successful today and how it must adapt or change to be successful tomorrow. Because we are a consumer society with rapidly changing technology and needs, it is insufficient for a manufacturer to be content with being the best at what it does—manufacturers must also consistently evaluate what else they should be doing. Changes in technology and what consumers may want or require will impact manufacturing in ways that are not immediately evident. Factories where learning is the norm will be in a much better position to rapidly identify impacts and respond competitively than factories where maintaining core competencies is viewed as a sufficient way of remaining competitive.

Guiding Vision

The second element Bowen et al. (1994) discuss is having a vision that guides decisions. It is likely that many readers will groan when they see the word *vision*. However, this overworked buzzword is absolutely critical to success. The vision does not need to be an elaborate description of what a manufacturing business intends to achieve, but, rather, a focal point that people can understand—an articulated idea of where the business is headed. The vision is what allows integration within the business. The vision provides the guide for developing capabilities, shaping the business strategy, and creating the culture necessary to support the vision. One excellent example of a vision is the Walt Disney Company's vision, as stated by Walt Disney.

> The idea of Disneyland is a simple one. It will be a place for people to find happiness and knowledge. It will be a place for parents and children to spend pleasant times in one another's company: a place for teachers and pupils to discover greater ways of understanding and education. Here the older generation can recapture the nostalgia of days gone by, and the younger generation

can savor the challenge of the future. Here will be the
wonders of Nature and Man for all to see and understand. Disneyland will be based upon and dedicated
to the ideals, the dreams and hard facts that have created America. And it will be uniquely equipped to
dramatize these dreams and facts and send them
forth as a source of courage and inspiration to all the
world.

Disneyland will be something of a fair, an exhibi-
tion, a playground, a community center, a museum of
living facts, and a showplace of beauty and magic. It
will be filled with the accomplishments, the joys and
hopes of the world we live in.

Those who have visited Disneyland or Disneyworld probably will
agree that the Disney vision is clearly reflected in those places. It took
years to build each of the entertainment/education sites, and yet the
vision developed by the Disney company guided it to its goals.

As with core competencies, visions must not be viewed as formal,
set-in-concrete statements. As a manufacturing business develops its
learning capabilities, gains knowledge, modifies its core competencies,
and understands more about its relationship with external events, it is
likely that its vision will also change. It is difficult to imagine a situation
in which the vision would not be refined over the passage of time. In a
learning factory, revising the vision to appropriately guide actions is
important.

A successful process for developing a vision is to include representatives from stakeholder groups, such as unions, supervisory personnel, support staff, shop floor workers, and midlevel managers, in the
development of the vision. One way of doing this is for top management to draft a vision statement, with stakeholder groups providing
review and modification of the vision. This exercise can have an added
benefit. Often a problem with vision statements is that they are lofty,
altruistic, and completely out of alignment with the operational values
of the business. When top management works with stakeholder
groups to develop the vision, such issues can be addressed. Where

there are critical conflicts or gaps, these can be acknowledged, and strategies for building a culture based on implementing the desired values can be developed. Readers can find additional information related to the development of a vision in chapters 2 and 10.

Leadership That Fits

The third element for a learning factory is operating with the organizational structure or pattern of leadership that is best for producing the work. The kind of work to be performed, the schedule for accomplishing the work, and factors such as whether a new product is being produced or a new process is being used will influence the best kind of leadership and the organization that is required to tackle the job.

A learning factory will have a system that can provide the information necessary for making decisions about the types of leadership, teams, and resources required to accomplish a specific kind of project. That system will be based on knowledge gleaned from previous experiences or from benchmarking efforts. For example, to market, design, develop, and produce a new product, a manufacturer may choose to use principles of concurrent engineering to form the team that will ensure the success of the product and that will get it to the marketplace ahead of the competition. Assembling a team of design engineers, material experts, production staff, as well as market and finance experts can offer the leverage a company needs to succeed in reducing design-to-supply time for new products. Boeing Corporation used such an approach in its highly successful production of the Boeing 777. One key factor in the rapid design, development, and production of the Boeing 777 was developing a team that used the advantages of concurrent engineering.

On the other hand, routine production requires a different type of leadership and management. It is a good candidate for utilizing a self-managed team that understands standard production processes, is accustomed to documenting exceptions to the processes, routinely uses lessons learned, and can augment the learning of the organization through such actions.

Ownership and Commitment

The ownership and commitment element concerns the feeling of ownership and pride that employees feel toward the work they are performing. Japanese factory owners depend on the loyalty that workers and management demonstrate toward each other. A generation ago, work in the United States was based on a similar norm. A person starting a job frequently remained with the same company for the duration of his or her work life. However, today it is unusual for a worker to remain with the same company for an entire work life. It isn't unusual for workers to make many employer changes and, often, career changes. This makes it difficult to develop a feeling of commitment and ownership among workers. One way factory owners can do this is through understanding and acknowledging the valuable resource that learning workers can be for the factory.

An example of where knowledgeable workers can make a significant difference is when factories undertake reengineering efforts in order to improve work processes, increase productivity, reduce costs, and achieve a competitive edge. Many businesses that undertake reengineering efforts turn to their top management and to outside consultants to conduct reengineering efforts. Our experience has been that frequently the workers (the real process owners) are the last to be consulted about how the work should be reengineered. In fact, sometimes they are never consulted. When this happens, a strong negative message is sent about what is expected from the workers. The idea that the worker is expendable and of little value may be the message that is sent. Factory owners who rely on their workforce for the information and ideas necessary for reengineering send a different kind of message. Their employees can sense that they are valued and that they are counted on for knowledge and information. It is more likely they will have a feeling of ownership for the work and commitment to the company than those workers whose knowledge and experience are ignored.

Pushing the Envelope

The fifth element suggested by Bowen et al. (1994) is the idea of constantly seeking improvements in products, processes, and capabilities. This involves more than simply trying to make small improvements—

the payoffs come when a stretch is involved. To push the envelope suc-cessfully, management must accept the value of intelligent risk taking. Workers must know that they are encouraged, in fact rewarded, for taking risks based on their knowledge, skills, and abilities. Whether a risk venture meets expectations, exceeds them, or fails miserably, the venture must be reviewed to capture the lessons learned. What made this successful; why were expectations exceeded; what happened, why did it fail? Learning organizations will build their own historical database of lessons learned, which, in turn, will guide new risk-taking ventures.

Sometimes a new way of performing a process or producing a product involves minor risk and might be viewed more appropriately as innovation, rather than a risky venture. Learning organizations encourage and support such innovation. Software development compa-nies are an example of an industry where innovative ideas and taking risks to develop products are encouraged. Factory owners can learn from such companies by studying the kinds of risks they take and how they use their workers to supply ideas for solving problems, generating ideas for new products, and taking intelligent risks. They can use their knowledgeable workers to become learning factories.

Prototypes

Using prototypes is the sixth element to consider for becoming a learn-ing factory. Prototypes are the models, mock-ups, and computer simu-lations of products and processes that help employees solve problems and learn faster and better. While the auto industry has a tradition of using model cars for testing design and production ideas, a Japanese manufacturer went beyond the model or prototype phase during the concept development phase of a luxury automobile intended for the U.S. marketplace. It sent a team of designers, developers, and mar-keters to live in luxury in the United States for several months. The objective of the team was to develop a thorough understanding of the expectations and lifestyles of those U.S. citizens who live in conserva-tive, very comfortable circumstances. As much as possible, team mem-bers were to "get inside the heads" of wealthy U.S. citizens in order to better understand preferences and expectations. They subsequently developed a prototype of the car that they later developed and very successfully marketed.

In addition to prototypes, which provide a way of pushing the envelope without putting the entire resources of the factory on the line, there are other, similar activities that factory owners can use for developing learning and knowledge. Garvin (1994) views traditional manufacturing projects as opportunities to develop smaller-scale experiments intended to produce incremental gains in knowledge. These experiments are usually associated with continuous improvement programs, which are common on shop floors. One example is the experimental efforts a team may make to reduce cycle time for manufacturing a product. Successful experiments can be reviewed to determine if similar efforts would benefit another part of the factory operations. Using ongoing projects as learning opportunities ensures a steady flow of new ideas. Using traditional projects as learning opportunities is more successful when the employee reward system encourages risk taking so that employees feel that it is okay to experiment; and it also requires that managers and employees are trained in skills necessary to perform and evaluate experiments. This evaluation permits concrete evidence of the success or failure of an innovation, and removes the subjectivity of gut-level feelings about the innovation.

In addition to ongoing projects, development projects offer an excellent opportunity for building new skills, gaining new knowledge, and creating new systems. Garvin (1994) explains that these are usually larger and more complex than experiments in ongoing projects. They are usually designed from scratch and take a clean-slate approach. Often they involve an entire site and affect the organizational structure of the site as part of the project development. These are good projects for integrating different functions and disciplines. General Motors' Saturn plant is an example of a development project that used innovation, new systems and knowledge, as well as untraditional ways of organizing the work and management. The production facility for the Saturn has served as a benchmark for others who want to make use of new ideas and knowledge. (Readers can find complementary information on strategic learning in chapter 7.)

Integration

The seventh and final element to consider for a road map for transitioning to a learning-based factory is integration. This occurs when the work to be performed and the individual tasks of the worker are aligned to accomplish the goals of the factory. In an integrated system, dysfunctional or unnecessary work is eliminated, tasks are reviewed to determine what value they add, and work is streamlined so that each task and each process contributes to building the desired product or providing a process in an effective, efficient manner. This is achieved through understanding the interdependence of tasks and how they must be integrated. Valuing teamwork and encouraging learning that results in applying knowledge from every member of the company helps a factory achieve integration. The goal of the team is to use members' collective intelligence to produce a product or process that is greater than that which any single member of the team could develop.

An example of how one functional organization used teamwork to successfully address complex problems is that of a contracts organization that was responsible for writing and letting large contracts for its company. The organization's internal customers, who depended on it for writing the contracts, frequently had serious complaints about its service. In addition, its measurable indicators revealed a regular pattern of poor performance. The contracts organization was under great pressure to dramatically improve its performance and gain customer favor. It accomplished this by working with a multidisciplinary team to identify the problems the organization was experiencing by determining its customer's requirements and by defining streamlined work processes. The team was made up from members of the contracts organization, experts from an operational effectiveness group, and members from a quality improvement group. The team was able to take a broad view of what the contracts organization needed to accomplish and how it needed to be integrated as part of a system for doing business in order to be successful. By looking at it as part of a total system, the team was able to consider and implement ideas that would have been deemed unrealistic if only an individual in the contracts organization was tackling the problem. Based on the results of the team's work, the contracts organization aggressively attacked its performance problems, significantly reduced customer complaints, and began functioning more efficiently.

Integration requires that companies respect their workers and that people listen to each other and give credence to thoughts, ideas, and questions of others. It requires an environment where employees are allowed to question the standard practices and make suggestions and recommendations for improved policies and standards. Again, infrastructures such as reward systems and performance-appraisal systems play a significant role in the degree of integration that can be achieved. If managers are rewarded for crosscutting functional boundaries to achieve goals and are not rewarded for building their own functional empires, then the attempts to integrate will be much more successful. Integration is key to efficiency and is what can ensure success of developmental teams.

The seven elements described here provide good information for manufacturers to consider as they transition to learning-based factories. Many manufacturers already incorporate some or all of the elements as part of the way they do business. For these manufacturers, the elements serve as criteria for evaluating where they are in their efforts to become or remain a learning organization.

The Learning Challenges for Learning Manufacturing Businesses

While the seven elements serve as a road map and provide evaluation criteria, manufacturers face other significant challenges in their quest to become learning organizations and develop knowledge workers—workers who continuously use the information they gather and skills they develop to improve their work processes and increase productivity. The seven elements (core capabilities, vision, leadership, commitment, pushing the envelope, prototypes, and integration) form critical infrastructures for the factory's learning organization, but the foundation for a successful learning organization is an educated workforce.

Earlier, the idea that education will become the business of private industry, rather than public/governmental responsibility, was mentioned. Many companies that are becoming learning- and knowledge-based are discovering that this is already true. They are assuming the role of basic educators for many of their workforce. This necessity is evidenced by Kaeter's (1993) report that 14 million U.S. workers read at or below the fourth-grade level. Arithmetic skills were found to be at a similar level. Kaeter states that, according to the Business Council for Effective Literacy, this is a level that is insufficient for 85 percent of the reading that is required on the job. In September 1993, a study conducted by Educational Testing Services of Princeton, New Jersey, and sponsored by the U.S. Department of Education revealed that of U.S. citizens 16 years and older, 47 percent demonstrate low literacy levels, although they describe themselves as reading and writing well or very well!

The study included five levels that described the respondents' skills levels. The first, level 1, represented 47 million U.S. citizens who were rated at the lowest level. These respondents had rudimentary reading and writing skills and were unable to locate a piece of information in a passage unless the wording was almost identical to the question. Level 1 respondents were also unable to locate an intersection on a map. Of these respondents, for 25 percent, English is a second language; 26 percent had physical, mental, or health conditions that impaired their ability to read and write; and 62 percent had not finished high school.

Approximately 50 million respondents were evaluated as being at level 2. These respondents were able to calculate the costs of purchases, but were unable to answer specific questions about newspaper articles. Respondents at levels 3, 4, and 5 had increasingly better skills and abilities. At levels 4 and 5, the estimate was that 40 million U.S. citizens could handle tasks including complex documents, frequent distractions, and information that required background knowledge.

The implication for a manufacturing business that wants to become a knowledge-based business is that there is a significant core of workers who currently are not capable of reading and writing in English, or of computing using mathematical formulas at a level that is sufficient for today's needs. Most manufacturers depend in some way on at least basic technology such as computers, fax machines, bar coding, electrical systems, and measuring and timing devices that require the ability to read, write, and/or compute. Manufacturers who are trying to implement the simplest of total quality measures, such as statistical process controls, depend on workers' abilities to collect data, perform mathematical calculations, prepare graphs and charts, and interpret and present data.

Companies that have faced the challenge of building a workforce with basic skills differ on the right approach to address the problem. Some companies have been advised that assessment tests are threatening to students and should be avoided. One such company, Onan (Kaeter 1993), agrees and avoids such assessment tests because it believes they are too frightening to adult learners. Instead, Onan designs courses that are flexible enough to handle a wide range of skills levels. When an assessment is unavoidable, Onan disguises it by testing everyone in a given job, thus removing the focus on a single individual.

Other companies, such as Magnavox (Ford 1992), use local community colleges to administer assessment tests to a volunteer sample of employees. After receiving the results, Magnavox had a sample of its work-related materials, such as manufacturing process instructions, engineering change notices, and blueprints evaluated by using the Fry readability formula and the Department of Defense's readability formula. Magnavox determined that its work materials were more complex than the average reading level of its workers and concluded that it needed to pursue a workplace literacy training program.

When a company determines that it does need to improve the literacy and mathematical skills of its employees, there are several ways to implement programs to accomplish this. Magnavox (Ford 1992) considered using private literacy firms, public adult education programs, community colleges, computer-based literacy programs, and federal funding. It opted to use federal funding and developed a program supported by the Charles D. Perkins Vocational Educational Act, which provides grants to states for adult training programs.

Most factories are located in areas where there are nearby community colleges, and often these colleges are willing to tailor or design courses to meet the needs of the factory. This is an excellent way to get the appropriate education for the factory workforce, as demonstrated by one company that provides its employees with basic problem-solving techniques by requiring that apprentices attend night school in addition to their daytime job training. The night school offers the place to perform and practice problem solving, and also provides students with a network that they can rely on for future problem solving and further learning (Roth et al. 1994).

In firms where English as a second language (ESL) is a problem, courses have been created to help workers gain the skills they need to function adequately in a learning organization. Companies that adopt some form of ESL training have experienced success and enjoyed increased productivity as a result of the improved communication skills of their workers (The Three R's 1993).

Obviously, manufacturers must be sensitive in the way they address illiteracy and innumeracy problems. The workers they have hired are for the most part intelligent, mature adults. Many of the workers are articulate and use good grammar and vocabulary in their oral communication. The breakdown occurs when reading and writing skills are required. Kaeter (1993) offers some suggestions for managing sensitive problems. One of the first recommendations is for manufacturers to recognize that a change in corporate culture will occur as a result of addressing illiteracy problems. One goal of the manufacturer will be to change from a place where management does the thinking and workers do the labor to one where management and workers both are expected to think. Workers will need to understand how their jobs are changing. They must understand that manufacturing strategies such as implementing SPC methods and increasing reliance on computers and robotics will change their jobs and what is expected of them.

Unions can be a great ally in helping workers accept the reality that they must improve their reading, writing, and computational skills in order to work in today's factory. The United Auto Workers has been at the forefront in backing basic education efforts. It has worked with automakers to create on-site education centers where factory workers can learn basic skills and take college preparatory classes. Factory management can take advantage of alliances with unions to establish an environment where it is acceptable, in fact desirable, to participate in skill-building courses that improve literacy and computational skills.

It is critical that any manufacturer that sponsors basic skills classes ensures that its trainers understand they are working with intelligent, responsible adults and that the material be geared toward adult interests. Often, people who are enrolled in a basic skills course have experienced years of humiliating failure in traditional classrooms and are fearful that they will have similar experiences in their factory-sponsored courses. They are embarrassed to be enrolled in the courses and feel that their peers and managers think they are dumb or incapable. Some companies have reduced the risk of such embarrassment by titling basic skills courses with names such as Shop Math, Technology Preparation, Preparation for Statistical Process Control, Company Procedures, or Business Forms (Kaeter 1993). Other companies have used titles such as Company Benefits as a way to encourage their ESL workers and other less literate workers to take a basic reading course.

The July 1994 issue of *Focus* (NCMS 1994) reports that the training methods for manufacturers are changing, offering more options to traditional classroom lecture-type learning. Electronic performance support systems and distance learning offer new training media in high-tech work environments. Interactive, electronically transmitted courses allow companies to merge classroom and shop floor so that more workers have access to a learning environment. Some manufacturers station computers with training and/or maintenance programs in the shop floor area. When workers experience downtime, they can use the computers for training exercises or to refresh their maintenance skills. Motorola University uses several methods for instructional delivery, including satellite, microwave, and video conferences. Most recently, manufacturers have begun to use CD-ROM as a means for increasing their instructional capability. These are excellent methods for other manufacturers who want to become learning organizations to adopt.

A final suggestion for making basic skills training acceptable in an organization is to cultivate the idea of training for all. In a factory where training is the norm, the stigma associated with being sent for training is removed. Providing people with the opportunities for training, sponsoring training on-site, arranging opportunities to share the information that is gained through training, and ensuring that people feel that it is okay to request training are essential to developing a learning organization.

Manufacturers also have opportunities to support knowledge and skills development beyond acquisition of basic skills. Many businesses pay for employee memberships in professional organizations and societies. This is an excellent way for employees to stay current in their field and also to keep looking ahead to future directions in their area of expertise. Manufacturers who support such activities are investing in learning opportunities for their employers and can expect to reap the benefits of this investment.

While many factories cross-train workers within a limited scope, such as in cellular manufacturing, workers who have an opportunity to work in many areas of their factory gain a broad, systemwide understanding of their environment and are able to provide fresh perspectives for solving problems and improving work processes. Workers with such a perspective can provide valuable insight and offer new directions and possibilities for the factory. Gaining such a whole-system perspective is an advantage that small and midsized manufacturers can offer their employees. This also is a good strategy for the learning factory.

Conclusion

The factory that demonstrates that it learns, so that it can develop knowledge to build the products and deliver the services that consumers want, will be a leading organization. When individuals understand that the corporation values them so much that it is eager to invest in their learning and that organization has the infrastructures that encourage teamwork and integration, then the individuals can share the information that becomes knowledge. It is important for manufacturers to remember that information, learning, and, subsequently, knowledge come from sharing on-the-job experiences, benchmarking other companies, formal classroom (or other media) experiences, and life experiences. To the degree that factory management encourages listening and learning from others, they have the opportunity to develop the knowledge that helps build the products and deliver the services that customers want. To the extent that an organization incorporates learning as a norm; understands the impacts of the external events on its environment; and has a workforce that is characterized by keen, eager minds, it will be able to develop necessary new core capabilities that enable the factory to rapidly meet changing consumer requirements. The organization will become a learning manufacturing organization that will be in a position to remain competitive and thrive.

References

Agyris, C., and D. A. Schon. 1987. *Organizational learning: A theory of action perspective.* Reading, Md.: Addison Wesley.

Bowen, H. K., K. B. Clark, C. A. Holloway, and S. C. Wheelwright. 1994. Development projects: The engine of renewal. *Harvard Business Review* 72 (5): 110–20.

Davis, S., and J. Botkin. 1994. The coming of knowledge-based business. *Harvard Business Review* 72 (5): 165–70.

Drucker, P. E. 1993. The new society of organizations. In *The learning imperative,* edited by R. Howard. Boston, Mass.: Harvard Business School Publishing.

Ford, D. J. 1992. The Magnavox experience. *Training and development* (November): 55–57.

Garvin, D. A. 1994. Building a learning organization. *Business Credit* (January): 19–28.

Kaeter, M. 1993. Basic self-esteem. *Training* (August): 31–35.

National Center for Manufacturing Sciences (NCMS). Strategic Development Staff. 1994. U.S. training methods benefit from injection of technology. *Focus* (July): 1.

Roth, A. V., A. S. Marucheck, A. Kemp, and D. Trimble. 1994. The knowledge factory for accelerated learning practices. *Planning Review* (May–June): 26–46.

Senge, P. 1990. *The fifth discipline.* New York: Doubleday.

Senge, P. M., R. Ross, B. Smith, C. Roberts, and A. Kleiner. 1994. *The fifth discipline fieldbook.* New York: Doubleday.

Stalk, G., Jr., P. Evans, and L. E. Shulman. 1993. Competing on capabilities: The new rules of corporate strategy. In *The learning imperative,* edited by R. Howard. Boston, Mass.: Harvard Business School Press.

The Three R's. 1993. *Training* (October): 64–65.

Additional Reading

Byrnes, J. L. S., and W. C. Copacino. 1990. Develop a powerful learning organization. *Transportation and Distribution* 31 (11): 22–25.

Peters, T. J. 1988. *Thriving on chaos.* New York: Alfred A. Knopf.

Reich, R. B. 1993. Strategic development staff of the National Center for Manufacturing Sciences. Training and education initiatives for smaller manufacturers are the new competitive mandate. *Focus* (July): 1.

Senge, P. 1990. The leader's new work: Building learning organizations. *Sloan Management Review* 32 (1): 7–23.

Shaw, R. B., and D. N. T. Perkins. 1992. Teaching organizations to learn: The power of productive failures. In *Organizational architecture,* edited by D. A. Nadler, M. S. Gerstein, R. B. Shaw, and Associates. San Francisco: Jossey-Bass.

CHAPTER 10

Management in the Agile Organization
Jennifer Macaulay

This book has focused on explaining how to create an agile environment in a small to medium-sized manufacturing company. The philosophy and techniques of agile manufacturing have been discussed in depth. It is time to turn our attention to the important job of managing for agile manufacturing. This chapter emphasizes the need for a new management style that supports an agile environment, the roles of a manager in an agile organization, and ways a manager can support the new agile culture.

Old Management Styles Will Not Do
Because an agile organization is based on implementing new ideas into the manufacturing processes, new ways of managing must be embraced to effectively support an agile environment. The old command-and-control style will not support the kind of employee empowerment and participation required for agile manufacturing.

As discussed in chapter 1, task alignment of support functions around the production system is vital to becoming agile. Management is another support function that must be aligned with the production system. This requires a paradigm shift on the part of many managers who think of themselves as controlling their employees. Agile organizations will require managers to understand and fulfill a primarily supportive role. Many organizations have begun representing this philosophy visually, with an organizational hierarchy shaped like an inverted pyramid (see Figure 10.1). This pyramid serves to remind employees that customers are the most important part of the organization. Shop floor employees, who create products for these customers, are the next most important. It is the job of support function employees and various layers of management to support these shop floor workers so that they can best meet the needs of the customers.

How does a manager support workers who can create improved processes and better meet customer needs? Managers must change their mind-sets. Managers in agile organizations do not rule a small turf or control their employees. Instead, they help their employees to create products that best meet the needs of customers. This means managers should help employees to better understand customer needs, redesign work processes, and continuously improve these processes. Supporting workers requires daily statements and actions to indicate that customers are the most important people in the organization. When the customer is considered the most important individual in the organization, supporting those workers who directly add value to the customer (through manufacturing goods or providing services) becomes

Figure 10.1. An example of an inverted pyramid organizational chart.

the priority of everyone in the organization. Such a radical shift in employee priority requires new roles of managers in agile organizations.

The Role of a Manager in an Agile Organization

Managers in an agile factory must play many roles: champion of the vision, team leader, coach, and analyzer of the business environment. In this section, we will discuss the tasks involved in filling each of these roles.

Vision Champion

Vision has been discussed extensively throughout this book. Managers are the people who communicate that vision, gain commitment to it, and translate it into the concrete work of teams. This occurs by translating the vision to strategy and the strategy to goals and actions. Chapter 7 deals extensively with this process.

In order for a vision to have an impact on an organization, it must be clearly understood. Communicating and clarifying the vision becomes an important task of the manager. Group members must understand the intent of the vision and what it will mean for them. The manager should spend time with the group helping members understand the overall purpose of the organization's vision and why it was selected. In a small manufacturing company, this might be accomplished through informal means, such as discussions around the lunch table or while "managing by walking around." As the company becomes larger, these informal methods are rarely sufficient. Communication about the vision needs to occur in company-wide meetings, team meetings, company newsletters, and other highly visible forums.

Once workers understand the organizational vision, the manager needs to help translate the organizational vision into a vision for his or her specific work group. This should be a creative process that includes all the work group members. When members help to create a group vision, they are more likely to clearly understand the implications of the vision, and they will more readily support it. A group vision should be aligned with the organizational vision. The organizational vision thus needs to be clearly defined to give good direction, but not too many constraints, to the group. When specific work group goals originate from it, the vision focuses and energizes the efforts of the work group. The vision acts to help workers understand how to prioritize their efforts. It tells workers what behaviors and activities are valued by the organization and will bring them rewards. By allowing groups to create their own vision, workers are given control over their work.

Belasco (1991) recommends the following steps in developing consensus around a group-specific vision.

1. If the work group is small enough, engage the entire group from the start of the process. If the group is too large (over six or seven members), the manager should share a draft version with a small group of trusted team members whose opinions are valued. Because of the confidentiality of this type of group, team members can test out rough ideas early in the process. As mentioned, the group's vision must be consistent with the larger organizational vision. In a larger organization where communication across groups is less frequent, it is important to include a few people from outside the work group. These people will help the team to gain perspective on how the group's vision aligns with that of other groups and with the organization as a whole.

2. The manager should share the vision informally with team members. The manager should ask people directly what they think. It is important for managers to stay flexible and not to get defensive, as the vision will go through many iterations. Presenting ideas informally and asking for feedback will reduce the perception that this is a final version of the vision.

3. Once the vision has taken a near-final form, the manager should hold a group meeting or meetings to discuss the vision and its implications. Feedback should, if appropriate, be incorporated into a final vision that has commitment from the group. Daily problems, decisions, and goals should be clearly tied to the vision.

The manager needs to make sure the group's vision incorporates the components of agile manufacturing discussed throughout this book. Important values include

- Customer focus
- Quality improvement focus

- Results focus

- Team focus

- Focus on employee empowerment

Customer focus. One of the hallmarks of an agile organization is that it puts the customer, and the customer's needs, first. Such a customer focus will not come about without work on the part of the manager. Managers will have to communicate and demonstrate the behaviors that put the customer first. When customer needs are well met, the organization will be more successful. Managers can ask workers to find ways of better understanding and meeting customer needs, and reward workers who demonstrate these behaviors.

Belasco (1991) recommends that one way to stress the importance of customer focus is to ask workers to spend time developing their relationships with both their internal and external customers. The development of strong customer–supplier communication is discussed in chapter 8. One method of developing a customer orientation for the group is through the development of customer–supplier contracts. These contracts detail what it is the supplier will strive to do for the customer. Customers and suppliers discuss how the supplier can best meet the needs of the customer and establish realistic, specific, and measurable goals for performance. The best way for the manager to stress the importance of customer relations is to measure customer–supplier contract performance. Managers should then base at least some portion of performance evaluation on how well these contracts are met.

Quality improvement focus. In order for an organization to become agile, it must find ways to improve its processes. Improvement is a continuing effort that must become part of the way people think about processes at work. It is the manager's job to foster an improvement focus for the work group. The manager is responsible for helping the workers understand the importance of continuous improvement. This requires the building of a climate that invites problem identification. This means that "shooting the messenger" is definitely not allowed. The messenger of bad news (the existence of a problem) is bringing the gift of an opportunity to improve the system. When workers understand that the identification of opportunities for process improvement are rewarded, rather than denied or blamed on workers, creative improvement begins to flourish. People should be encouraged to bring forward problems, even if they have not yet thought of a solution. When people feel they must know the solution before they can bring forward the problem, they tend to keep problems to themselves. Many minds working together to solve a problem can almost always come up with a better solution than one person working alone.

Nothing fosters an improvement focus like success with improvement efforts. As was mentioned in chapter 4, quick wins help sell IWPR efforts to the entire organization. Managers must encourage the team members to identify small improvements they can make immediately. These quick wins will help people feel that they are empowered to make changes in their work lives and will motivate other improvement ideas.

Results focus. A results focus means that outcomes, rather than activity, are valued. Many organizations measure and reward the amount of activity that takes place, rather than the results that come from the activity. For example, some quality improvement efforts have failed because they have focused on the number of quality improvement teams trained and working on problems, rather than the improvements the quality teams have actually made. When an organization emphasizes activity, instead of results, it frequently has a lot of very busy people who never seem to get much done.

One way to encourage a results focus is to reduce fear of failure. In their book, *Organizational Architecture,* Nadler, Gerstein, and Shaw (1992) emphasize the importance of reducing risk-averse behavior to gain a results focus. When people perceive a "high risk-to-reward ratio" (they believe the risk far outweighs any potential rewards), they become activity oriented and fail to attempt new, potentially effective solutions. To reduce risk-averseness, it is important to reward prudent risk taking, even when ideas are not successful. When people are afraid to fail, they cannot be creative. As is discussed in chapter 9, learning from failure can frequently lead to long-term success.

In order to best capitalize on learning experiences, it is important to evaluate all projects and to communicate lessons learned to others in the group and within the organization. Institutionalizing a lessons-learned process goes a long way in stressing the importance of continuous improvement. For example, a manufacturing company could institutionalize a process of organizational learning by asking shop floor or distribution staff to keep a daily log of problems or information learned. Managers could ask employees to share and discuss those problems and learnings at regular team meetings. Managers would then be responsible for sharing this information with other managers and passing new information back to their own teams. In this way, problems or learnings from any employee help the entire staff to learn. The larger the company, the more formalized this process should be to make sure all parts of the organization are communicating.

Team focus. The manufacturing workplace has become more and more complex. It has reached a stage where it is no longer feasible to expect one person to know enough about all parts of a task to be able to work alone. Cross-functional teams with individuals who bring diverse skills and talents to the group are needed to implement agile manufacturing techniques. Unfortunately, most people raised in the United States have not had much experience working in teams outside of athletics. Most organizations in the United States value individualism and reward individual effort. This means managers in agile organizations will have to teach their workers team skills and the value of team efforts. As was mentioned in chapter 4, important team skills that should be developed are interpersonal communication, conflict management, meeting management and facilitation, planning, and coordinating. *The Team Handbook* (Scholtes 1988) is an excellent resource for managers who need to develop team skills for themselves and their group.

The manager in an agile organization will have to foster an environment that is supportive of a team focus. In their excellent book *Leading Teams: Mastering the New Role,* Zengler et al. (1994) recommend to managers the following daily activities to demonstrate dedication to a team focus. Team managers should

- Develop employees' ability to make decisions that affect their work.

- Allow the team to solve its own problems without intervening.
- Avoid taking back responsibilities that have been given to a team.
- Obtain the developmental resources people need to perform to their potential.
- Improve their own skills.
- Allow others to express their opinions before they do.
- Relinquish the perks that separate managers from their teams.
- Recognize and celebrate progress toward team goals.
- Change systems and processes that do not support a team focus.

Leadership behavior that reflects a team focus must be demonstrated every single day. It may take some time and experience with the benefits of teams before workers become converted to a team focus.

Empowerment. Empowerment is defined as the process of enabling people to do what they are capable of doing and for which they are held responsible. Managers must encourage others to be involved in decision making, improvement efforts, and other job-related procedures to generate increased buy-in. An empowered organization has pushed control and decision making down to the workers, freeing up managers to support these people and analyze the business environment of the organization.

Empowerment is important because it leads to ownership and creative involvement in improvement efforts. Empowered employees are the right people to be making the decisions that affect their work processes because they have the most knowledge about the systems. Empowerment increases commitment to creating more efficient and effective processes. Efficient and effective processes lead to organizational success.

Empowering workers can be difficult as a manufacturing company grows from a small organization of a handful of people to hundreds. In the small company with only a few managers, frequently owners or founders, managers may feel uncomfortable relinquishing some control over decisions. These managers must recognize that, as the company grows, their valuable time can be better spent on strategic issues and long-term vision than on day-to-day decision making. By clearly communicating the vision and strategic direction of the organization they can guide daily activities without having to be directly involved. Also, those workers who most closely work with a process are in a much better position to make accurate decisions about how to improve it than are distant managers.

An example of effective empowerment in agile manufacturing is allowing the workers of the process to make decisions about how to best redesign that process to make it more efficient. The shop floor worker who frequently must scrap a product due to a high defect rate in certain supplied parts is in a much better position to choose high-quality suppliers than the manager who never directly works with those parts.

In order for agile manufacturing to come alive, employees must feel empowered to make decisions and take responsibility for improving systems. Empowerment requires the unleashing of human potential. Most of the workforce in the United States today has grown up in an environment with little or no empowerment. Workers do not expect or understand how to take more responsibility. It is the manager's job to help them become empowered. Nadler et al. (1992) suggest that in order to empower workers, the following conditions must be met.

- A vision and strategy clarify priorities for action.
- A "vital few priorities" provide real guidance.
- A high "reward-to-risk ratio" encourages a results focus.
- Accountabilities are clear and motivate performance.
- A leader exercises "minimum acceptable control" over decisions.
- Resources are available to support activities.

Nadler et al. go on to discuss the following high-leverage actions that support empowerment.

- Managers should form groups that are small and free to make their own decisions.
- Managers can replace rules and regulations with general guidelines that clarify and reinforce the vision.

- People must be held consistently accountable for their successes and failures.
- Failure should be treated as a learning opportunity.
- Managers must support the training and education that enable people to take on more responsibility.

Empowering workers means managers must accept letting go of control. Most managers have become successful in their lives by being responsible and doing good work. Now, managers are being asked to let others be responsible and do good work. This leap of faith requires an assumption about workers: that they *can* and *will* do a good job. Managers need to give their employees the appropriate skills training so that they *can* do a good job and consistently hold workers accountable for their work so they *will* do a good job. Developing people is a vital part of empowering them.

Team Leader

The manager in an agile organization must also play the role of a team leader. This role has several components: facilitator, organizational boundary manager, and educator. Each of these roles is discussed here.

Facilitator. A facilitator is someone who assists with the interpersonal components of teamwork. As a team facilitator, it is a manager's job to aid the interpersonal interactions among team members. All teams will need a person to tend to the "softer" side of what is happening when they work together. Excellent facilitators recognize when non–task-related factors are keeping a group from being productive. Such non–task-related factors might be fatigue, conflict, an overbearing team member, or lack of group focus. Facilitators must be aware of group dynamics and keep the team as productive as possible. The role of facilitator demands strong interpersonal skills to keep a team of individuals communicating and working together. It is highly recommended that managers develop their own communication and conflict management skills.

Also, as facilitator, it is important that managers help the team understand its goals and stay on track. This does not mean imposing goals or constraints, but rather helping the team set its own goals and identify the constraints under which it must work.

Organizational boundary manager. It is the job of the manager of a team that is part of a larger organization to manage the interactions with other parts of the organization. The boundary manager manages the interface with other teams so that the work flows smoothly between teams and helps acquire needed resources for the team. In small organizations, boundary management is much simpler because most people cross several parts of the organization, or staff size is small enough that everyone is aware of what others in the organization are doing. However, once the organization gets large enough to have employees who do not regularly interact with employees from other parts of the organization, boundary management becomes an important communication and interaction task of the manager.

In a traditional organization, managers frequently attempt to build and protect turfs. By engaging in turf battles, managers in traditional organizations attempt to maximize their group's resources and success, but sacrifice the overall success of the organization. In an agile organization, where maximally effective improvement solutions require input from all parts of the organization, boundary management is especially important. Managers must encourage communication and resource sharing across boundaries to increase the effectiveness of the organization. It is especially important when an innovative new process is being developed in one group. It is crucial that resistance from the larger organization not undermine changes. Communicating with all people in the organization who might be affected by the change cannot be stressed enough.

It is also the manager's job as boundary manager to help the team understand what is going on in the larger organization. This is especially true in larger companies. Employees need to be kept informed of the progress of other groups and the changes in the organization as a whole. Organizational communication needs to be facilitated so employees see the big picture and understand how their work is supportive of larger organizational efforts. This is especially true of non–production-oriented support function staff who may not see how their activities add value to the manufactured goods.

Educator. The importance of education to an agile organization was discussed in depth in the previous chapter. The manager acts as the person in the organization who makes sure that each individual is getting the right training and education to meet his or her needs. Education and training can be obtained and provided in many different ways. One responsibility of the manager is to obtain the needed funds to allow individuals to gain developmental opportunities. Smaller manufacturing companies frequently do not have large training budgets on which to rely. Low-cost developmental opportunities can be created by managers who call upon employees to help teach, mentor, or guide team members through new skill-building experiences. Managers can act as trainers themselves, either informally on the job or through stand-up training. Team members should be encouraged to share new skills and concepts with one another. Further, the manager develops the skills of his or her workers by modeling the appropriate skills him or herself.

Coach

The third role of a manager in an agile organization is that of coach. The best way to understand the role of coach is by analogy. An athletic coach exemplifies what is required in coaching in the workplace. First, the coach's job is to help the athlete (worker) to perform to his or her potential. The coach's job is to support that person as best he or she can. The second function of a coach is to teach new skills and to deliver feedback about performance. People are a coach's most important resource. They must be developed to their fullest potential. Third, coaches are evaluated by how well their athletes (workers) perform. It is not their job to sink baskets themselves, but to help their workers sink baskets. The same is true of managers.

The manager's role as coach will require him or her to provide ongoing feedback and developmental support to team members. Coaching requires an investment of time and frequently other resources, but this investment will ensure the continued growth and improvement of workers. Coaching should include providing encouragement, performance feedback, and specific performance improvement suggestions. Further, a coach should assist workers with developmental goal setting and career planning.

In *Inspiring People at Work* by Thomas Quick (1986), a distinction is made between long-term and short-term coaching. Long-term coaching involves developing the skills the team members need to continue to grow and be successful in the organization. As stated in the preface, the quality of the organization's employees—the scope and depth of their knowledge, their ability to solve problems, and their ability to learn and absorb new methods of technology—is a key aspect of becoming and remaining agile. Long-term coaching should become a formalized process in each group. Several times each year, managers should discuss with all team members their hopes, dreams, and goals for themselves in the organization. Developmental goals should be set with each person, and the manager should assist the person in getting needed resources for reaching these goals. Career planning can be difficult in small manufacturing companies with little room for promotion. Job expansion, cross-training, and empowerment are especially important in these organizations to keep employees growing and challenged.

Short-term coaching is centered around deficiencies and helps employees perform better today. The following guidelines are meant to assist in effective problem-centered coaching (Quick 1986).

- Agree on the problem. Both the manager and the employee must agree that standards are not being met.

- Describe alternative solutions. Together the manager and employee should generate several options for how to solve the problem.

- Determine the goal. Both the manager and the employee must agree on what goal behavior will look like. It should be described in a way that is specific and measurable. Strategies and tactics needed to obtain the goal should be described.

- Have a backup plan. The world is always changing; therefore, a second plan is needed if the first does not work. Manager and employee should determine how and when to regroup to establish the backup plan.

- Define the manager's responsibilities. How can the manager help? Determine if the manager should obtain needed training, provide on-the-job assistance, arrange for a mentor, or just stay out of the way.

- Outline an oversight process. Determine how and when the manager and employee will review progress toward the goal. Weekly? As needed? When asked?

- Provide ongoing coaching. Some tips for giving positive and negative feedback are in Table 10.1.

Both long-term and short-term coaching will help employees develop the skills and knowledge they will need to be successful in the workplace. Remember, as a coach, a manager's success is based solely on his or her employees' success.

Business Analyzer

The fourth and final key role of the manager in an agile organization is that of business analyzer. The role of business analyzer involves monitoring and acting on changes in the business environment. In chapter 2, a model for effectively managing change is laid out. The first step in this model is to diagnose the business situation. As is stressed in chapter 7, it is the manager's role to model both strategic thinking and strategic learning behavior. Both of these behaviors require understanding the business environment. Thus, the manager's role as business analyzer is key in setting strategic direction and effectively managing change.

As business analyzer, it is the manager's responsibility to gather information about the group's, organization's, and industry's environment and share this with the group. This information needs to be analyzed

Table 10.1. Hints for providing constructive positive and negative feedback.

Positive feedback	Negative feedback
1. Be consistent.	1. Describe the behavior, not the attributions.
2. Keep it in perspective.	2. Get agreement on behavior and standards.
3. Make it timely.	3. Listen to the other side of the story.
4. Be specific.	4. Emphasize improvement techniques.
5. Do not be stingy.	5. Set goals for improvement.
6. Make it public.	6. Make it private.

to determine likely impacts on the group or the organization. How are changes likely to influence the group? Are new technologies emerging? Are customer needs changing? Does a competitor have a new product? It is the business analyzer's job to determine how the world is changing around the team and to keep team members informed of these changes and their impacts.

In an agile organization, continuous improvement demands active business analysis. It is important to understand that unless managers fully empower employees, it will be difficult to find the time to adequately analyze the business environment. When managers empower workers to make their own decisions and be accountable for their own work, managers receive in return the extra time needed to fulfill this crucial task. Managers do not have to spend large amounts of time analyzing the environment. Spending small amounts of time each week gathering business environment data will help keep the effort from becoming overwhelming, as well as helping to keep managers current. Belonging to and attending industry professional groups can help keep a manager on top of new developments. Reading national or international business or industry journals can also keep managers abreast of trends in the industry. Spending time each month in contact with customers is the best way to gain important knowledge about customer needs, complaints, or ideas. Both managers and team members should try to expose themselves to customer feedback on a regular basis.

Supporting the New Culture

There are two crucial ways to support the new agile culture. The first is through the velvet glove mandate discussed in chapter 2. The velvet glove mandate basically says that if managers are not supportive of the new culture, they need to be removed from their position and relocated to a nonmanagement position or let go. Changing to an agile organization will require an investment of time, money, and energy. An organization simply cannot afford to have managers undermining efforts at becoming agile. This policy must be made abundantly clear right up front, and it must be followed strictly, especially in smaller companies where interaction among all members of the organization is frequent. If managers or employees see other managers undermining the effort and getting away with it, the change to an agile organization will not happen.

Once the new culture is in place, the velvet glove mandate should apply to employees as well. Employees who will not participate in the new practices will not be productive in the organization. These individuals need to be let go. If other employees see an employee not participating in the new techniques and processes, and not being sanctioned for this behavior, the importance of agile behavior will be undermined. It cannot be said strongly enough—*give people fair warning, but act on it!*

The second way to support the new culture is to align the infrastructure with the new culture. Taking a systems approach to change has been emphasized throughout this book. Managers in small to medium-sized manufacturing companies frequently have quite a bit of influence over many of these systems. This includes such systems as performance appraisal, selection, reward and recognition, organizational performance measures, and organizational learning. Each of these is discussed in turn.

Performance Appraisal Systems

Performance appraisal systems are one of the most powerful tools for changing organizational behaviors. These systems outline what behavior is valued, rewarded, and punished. Unfortunately, many organizations have spent little time developing a performance appraisal system that adequately communicates desired behavior and performance standards. This is especially true in very small organizations that tend to have no formal performance appraisal system. Frequently, behavior that is espoused as desired is not what is measured or rewarded in the organization. For example, quality may get lip service, but production speed is what is measured and rewarded.

Performance must be measured and rewarded according to the vision. It should include many of the customer-oriented measures discussed in chapter 8, such as how well a manager or employee is succeeding in meeting the needs of his or her customers, working on cross-functional teams to solve problems, and creatively improving systems. It does not make sense to expect new behavior from people if they are not evaluated based on these behaviors. All performance review systems must measure and reward the desired new behaviors. Managers should measure team members based on both organization and team visions.

One effective way to evaluate performance in a team-oriented culture is to train teams to do their own performance appraisals. Team members receive feedback on their performance from their teammates (including the manager). Many workers find this information to be more specific, more fair, and more useful than being evaluated by a manager with whom they may not have spent a great deal of time working. If a manager decides to have team members conduct their own performance appraisals, team members will need the appropriate training. Include training on the performance appraisal system, as well as skills and practice in delivering constructive feedback.

A recent article by Spink (1994) provides an example of a manufacturer doing team performance appraisal. Quickie Designs manufactures lightweight wheelchairs. Quickie has phased in team performance appraisal over the past few years. When it first started this team performance appraisal, the individual being evaluated and one or two peers would provide performance appraisal information to the manager. This information was integrated by the manager and fed back to the individual being evaluated. Now Quickie does team performance appraisals in a full-team meeting at which each team member is appraised by all other team members in turn. Employees at Quickie report that they feel the new performance appraisal system is much more helpful and developmental, as opposed to evaluative. They feel their team has become more cohesive, and team members feel more ownership in improving.

Finally, the manager may want to consider evaluating the team as a whole. How do members work together, coordinate efforts, and communicate? Evaluating the team as a whole underscores the importance of cooperation and reduces intergroup competition. *The Team Handbook* (Scholtes 1988) is an excellent resource for tools for developing a team assessment instrument for the group.

Selection Systems

The team performance appraisal is one way to support an agile culture. Teams also can do a good job of selecting new members and making promotion decisions. Team members better understand the skills they will need in a new member. They also can best evaluate how well they will be able to work with a new team member. Once teams begin to form, members can become quite close. It is useful to allow team members to select their own new member, as it tends to increase acceptance of the new member.

Again, just as with evaluating performance, it is crucial that new members be selected into the organization based on how well they fit into the new vision and culture. The manager will want to be sure that members are selected who indicate a belief in the customer, results, the team, and quality improvement. Individuals should be selected who demonstrate the ability to easily learn new information. This is especially important in a constantly changing agile organization. Finally, individuals should be selected who demonstrate excellent interpersonal skills. Interpersonal skills are vital to good teamwork and are frequently more difficult to teach than other, less "soft," skills.

Similarly, those individuals who embody the vision should be promoted. This is the ultimate reward for desired behavior. An individual who is an excellent individual contributor but does not exhibit team, customer, and improvement focus should not be promoted.

Reward and Recognition Systems

Once again, a manager must reward those new behaviors that the organization desires. If a manager wants to see cross-functional team cooperation, he or she cannot reward on a competitive, individual basis. If a manager wants people to be customer oriented, the people should be rewarded on the basis of customer satisfaction. It is very important to reward based on results, not activity. In their book, *The Race without a Finish Line*, Schmidt and Finnigan (1992) recommend the following steps in rewarding and recognizing good performance.

- Publicize the good, not the bad. Managers need to catch people doing things right, and celebrate!

- Publicly reward. Other people should see who and what behavior is being rewarded. It not only makes the person being recognized feel good, it gives people an example of desired behavior.

- Give personalized rewards. Managers need to show that they care about the person. Rewards are more motivating when they are things people really want. A manager's employees are his or her customers, so the manager should find out what they want and deliver it.

- Make rewards timely. In rewards (as in comedy), timing is everything. The person should be rewarded while the behavior is still fresh in his or her mind.

- Specify why a person is rewarded. Everyone should understand why the individual is being rewarded. This models the desired behavior and associated rewards for others, and increases the motivation to perform the desired behavior.

- Encourage others to reward. Managers are not the only ones who can hand out praise. Team members should be encouraged to recognize one another.

- Reward the team. When an entire team has worked hard to improve a system, a process, or a product, all team members should be rewarded. Many agile organizations use profit sharing systems such as the Scanlon plan (Perry 1988) to reward improved organizational performance. The Scanlon plan is a group and individual incentive program built around employee participation in determining production goals.

Organizational Measurement

Organizational measurement is covered extensively in chapter 8. It is mentioned again here because it is an important component of what a manager needs to communicate with his or her team. Managers must help their team understand how they contribute to the success of the organization. Measures of team performance must link closely with measures of organizational performance.

In small manufacturing companies, the head of the company may be able to consistently communicate this information. However, in larger organizations this communication should be formalized and passed down to the managers of different teams.

Organizational Learning Processes

As has been mentioned in this chapter and in chapter 9, in an agile organization, lessons learned in one project or team must be shared with the entire organization. If each group can learn from those that have come before it, the group will be able to work more efficiently and more effectively. Organizational learning can help teams to work faster and smarter than before, thus making the organization and its products more competitive.

For organizational learning to take place effectively, processes must be put into place that guarantee that information will be communicated throughout the organization. In small companies it is easier for this information to be passed along on an informal basis. However, as the company becomes larger, formal systems for recording and communicating lessons learned become necessary. Any system that makes evaluation of projects and processes a mandatory part of the job will facilitate organizational learning. Systems for communicating lessons learned could be accomplished via company meetings, frequent refresher training, company-wide newsletters, bulletin boards, or computer networks. Management should choose a method that works well for their organization. The smaller the company, the more informal these processes may be.

Managers should create an environment in their team where the frequent exchanges of lessons learned are an expected part of the job. People should be asked to share stories over coffee in the lunchroom or whenever a new group member is introduced.

Summary

Managers in agile manufacturing companies will face many challenges. The manufacturing and support processes will be changing rapidly and frequently. Both employees and management are being asked to behave differently and to develop new skills. There is a lot of work for managers to do. The major challenges are new roles to fill as manager and new systems to develop in order to support the new culture.

In their recent book *The Race Without a Finish Line*, Schmidt and Finnigan (1992) discuss lessons learned from Malcolm Baldrige National Quality Award winners. They describe the critical behaviors of the new manager as follows:

- Give customer needs top priority.
- Empower and develop workers.
- Encourage continuous improvement.
- Encourage problem identification and problem solving.
- Champion the vision in word and deed.
- Make systems supportive of agile practices.
- Reward team effort.

Managers of this type are needed for agile manufacturing to thrive. When managers can give up old behaviors and begin to learn the skills necessary to behave in a new, supportive way, employees become empowered to find creative ways to make processes more effective and efficient.

Management plays a very important role in agile organizations. Although they empower their people to do much of the old work of management, managers will find themselves very busy with a new set of roles and responsibilities. The most important role of a manager in an agile organization is to be a role model. Managers must help their employees understand their new roles, their new tasks, and a new way of thinking.

References

Belasco, J. 1991. *Teaching the elephant to dance: The manager's guide to empowering change.* New York: Plume.

Nadler, D., M. Gerstein, and R. Shaw. 1992. *Organizational architecture: Designs for changing organizations.* San Francisco: Jossey-Bass.

Perry, N. 1988. Here come richer, riskier pay plans. *Fortune* 118 (14): 50–58.

Quick, T. 1986. *Inspiring people at work.* New York: Executive Enterprises.

Schmidt, W., and J. Finnigan. 1992. *The race without a finish line: America's quest for total quality.* San Francisco: Jossey-Bass.

Scholtes, P. 1988. *The team handbook.* Madison, Wis.: Joiner Associates.

Spink, P. 1994. Team directed management. *Target* 10:18–22.

Zengler, J., E. Musselwhite, K. Hurson, and C. Perrin. 1994. *Leading teams: Mastering the new role.* Homewood, Ill.: Business One Irwin.

APPENDIX A

Author Biographies

Lawrence O. Levine, editor

Mr. Levine is a senior research engineer at Pacific Northwest National Laboratory (PNNL) whose professional experience has included management systems assessment, application software design and development, technology assessment, and R&D planning and assessment. His recent focus has been on improving operational effectiveness, particularly reducing cycle time in administrative processes and implementing TQM in R&D organizations. Mr. Levine has a B.S. in engineering from the University of Michigan and an M.S. in industrial administration from Carnegie Mellon University. He is a member of the Institute of Industrial Engineering, Association for Manufacturing Excellence, American Production and Inventory Control Society, and National Council of Systems Engineering.

Joseph C. Montgomery, editor

Dr. Montgomery is a senior research scientist at PNNL. His primary areas of research and consulting include organizational change management and organizational design, especially as linked with business process reengineering and process improvement efforts. Other areas of expertise include organization development, organizational performance measurement, and strategic planning. Dr. Montgomery received

285

a Ph.D. in industrial/organizational psychology from Colorado State University and a B.S. in physics from the University of Washington. He is an adjunct faculty member at Washington State University, Tri-Cities, and is a member of the American Psychological Association, Society of Industrial/Organizational Psychologists, and the Academy of Management.

Monty L. Carson

Mr. Carson, a research engineer with PNNL, is involved in a number of manufacturing and business process improvement and redesign projects. Before joining Battelle, Mr. Carson worked in a variety of engineering roles for Honeywell, Industrial Automation Division; Allied-Signal Aerospace Company, Garrett Engine Division; and Motorola, Government Electronics Group. He holds M.S. and B.S. degrees in industrial engineering from Arizona State University and is a member of the Institute of Industrial Engineers.

Cody J. Hostick

As a senior research engineer, Mr. Hostick has been primarily involved in the design and assessment of manufacturing systems and manufacturing support functions. Currently, he is supporting multiple small manufacturing firms as part of the PNNL's small business assistance effort. Mr. Hostick has a B.S. in industrial engineering from Oregon State University. He is a licensed professional engineer in mechanical engineering and is certified by the American Production and Inventory Control Society in production and inventory management.

Jennifer L. Macaulay

Dr. Macaulay is a research scientist with the Battelle Seattle office. She has been involved in research and consulting in the areas of business process redesign, organizational assessment, leadership development, change management, and strategic planning. She also develops and conducts training in management and leadership effectiveness. Dr. Macaulay has an M.S. in psychology from Western Washington University and a Ph.D. in organizational psychology from the University of Washington.

Brian K. Paul

Dr. Paul is an assistant professor of industrial engineering at Oregon State University in Corvallis, Oregon. He formerly served as a senior research engineer at Battelle Pacific Northwest National Laboratories in manufacturing production systems analysis and design. Before working at Battelle, Dr. Paul also worked as a manufacturing engineering consultant, quality engineer, and industrial engineer for the McDonnell Douglas Helicopter Company, Honeywell Industrial Automation Systems Division, and Boeing Military Airplane Company, respectively. He received his Ph.D. in industrial engineering from the Pennsylvania State University, his M.S. in industrial engineering from Arizona State University, and his B.S. in industrial engineering from Wichita State University. He holds memberships in the American Powder Metallurgy Institute, Society of Manufacturing Engineers, and Institute of Industrial Engineers.

Linda R. Pond

Ms. Pond is a technical group manager within Battelle's Operational Improvement Group, Technology Management and Social Sciences Department. In addition to her line management role at PNNL, she has worked in the areas of developing learning organizations and organization development. She holds a B.S. in organizational communications from the University of Central Florida and an M.S. in human resource management from Florida Institute of Technology.

Russ E. Rhoads

Mr. Rhoads, as a senior program manager at Battelle, has held a variety of line management and program management positions since coming to PNNL in 1975 and has contributed to many large, interdisciplinary research programs. He has been involved in developing an integrated, systems approach to performance enhancement and is currently leading the laboratory's programs to use this approach to improve the operating performance of industrial facilities in the federal government. Mr. Rhoads holds a B.S. in physics and an M.S. in nuclear engineering from the University of Washington.

David J. Lemak

Dr. Lemak is an assistant professor of management at Washington State University, Tri-Cities, who has worked closely with PNNL in a variety of research and consulting projects. His teaching responsibilities include graduate and undergraduate courses in strategy, organizational theory, and general management. He is a former department head for Management and Organization at the U.S. Air Force Academy. He received his Ph.D. in strategic management from Arizona State University, an M.B.A. from Indiana University, and a B.S. in political science from Ohio Wesleyan University. Dr. Lemak is a member of the Academy of Management and the Decision Sciences Institute.

APPENDIX B

List of MEP Centers

The National Institute of Standards and Technology in Gaithersburg, Maryland, has partnered with local and state governments across the country to deliver manufacturing technology services to small manufacturers via the Manufacturing Extension Partnership (MEP). Forty-four MEP centers currently exist, with services ranging from technology planning to training to business consultation. These 44 centers are listed in this appendix. This list is growing and subject to change. For more information concerning MEP services and center locations, contact the following.

The Manufacturing Extension Partnership Office
The National Institute of Standards and Technology
Building 224, Room B115
Gaithersburg, MD 20899-0001
Telephone: 301-975-5020
Fax: 301-963-6556
E-mail: MEPinfo@micf.nist.gov

Arizona Applied Manufacturing Center
Phoenix, Arizona
602-392-5184

California Manufacturing Technology Center
Hawthorne, California
310-355-3077

Pollution Prevention Center
Santa Monica, California
310-453-0450

Connecticut State Technology Extension Program
Storrs and New Britain, Connecticut
203-486-2585

Delaware Manufacturing Alliance
Newark, Delaware
302-452-2520

Georgia Manufacturing Extension Alliance
Atlanta, Georgia
404-894-8989

Chicago Manufacturing Technology Extension Center
Chicago, Illinois
312-265-2020

Iowa Manufacturing Technology Center
Ankeny, Iowa
515-965-7040

Mid-America Manufacturing Technology Center (MAMTC)
Overland Park, Kansas
913-649-4333

>   MAMTC Colorado Regional Office
>   Fort Collins, Colorado

>   MAMTC St. Louis Regional Office
>   St. Louis, Missouri

>   MAMTC Southern Missouri Regional Office
>   Rolla, Missouri

MAMTC Wyoming Regional Office
Laramie, Wyoming

Kentucky Technology Service
Lexington, Kentucky
606-252-7801

Maine Manufacturing Extension Partnership
Augusta, Maine
207-621-6350

Maryland Manufacturing Modernization Network
Baltimore, Maryland
410-333-6990

Massachusetts Manufacturing Partnership
Boston, Massachusetts
617-333-6990

Michigan Manufacturing Technology Center
Ann Arbor, Michigan
800-292-4484

Minnesota Manufacturing Technology Center
Minneapolis, Minnesota
612-338-7722

Nebraska Industrial Competitiveness Service
Lincoln, Nebraska
402-471-3769

New Mexico Manufacturing Extension Program
Albuquerque, New Mexico
505-272-7800

New York Manufacturing Extension Partnership
Troy, New York
518-283-1010

Hudson Valley Manufacturing Outreach Center
Fishkill, New York
914-896-6934

Manufacturing Outreach Center
Binghamton, New York
607-774-0022

New York City Manufacturing Outreach Center
New York, New York
212-240-6920

Western New York Manufacturing Outreach Center
Amherst, New York
716-636-3626

North Carolina Manufacturing Extension Partnership
Raleigh, North Carolina
919-515-5408

North Dakota Manufacturing Extension Center
Bismarck, North Dakota
701-328-5300

Great Lakes Manufacturing Technology Center
Cleveland, Ohio
216-432-5324

Miami Valley Manufacturing Extension Center
Kettering, Ohio
513-259-1340

Lake Erie Manufacturing Extension Partnership
Toledo, Ohio
419-531-8610

Plastics Technology Deployment Center
Cleveland, Ohio
216-432-5300

Oklahoma Industrial Extension System
Tulsa, Oklahoma
918-592-0722

North/East Pennsylvania Manufacturing Extension Partnership
Bethlehem, Pennsylvania
610-758-5599

Southeastern Pennsylvania Manufacturing Extension Partnership
Philadelphia, Pennsylvania
215-464-8550

Southwestern Pennsylvania Industrial Resource Center
Pittsburgh, Pennsylvania
412-687-0200

Southeast Manufacturing Technology Center
Columbia, South Carolina
803-252-6976

Tennessee Manufacturing Extension Partnership
Nashville, Tennessee
615-741-2994

Texas Manufacturing Assistance Center
Austin, Texas
512-936-0235

A. L. Philpott Manufacturing Center
Martinsville, Virginia
703-666-8890

Virginia Alliance for Manufacturing Competitiveness
Richmond, Virginia
804-786-3501

Washington Alliance for Manufacturing
Seattle, Washington
206-622-3456

West Virginia Partnership for Industrial Modernization
Huntington, West Virginia
304-696-4852

Northwest Wisconsin Manufacturing Outreach Center
Menomonie, Wisconsin
715-232-2397

Index

triple diagonal (TD) modeling—*continued*
  example, 82–85
  flows in, 82, 83, 86, 87–88, 90
  level of detail, 86
  processes in, 82, 83, 86
  scope, 85–86
  steps, 85–90
  team, 85, 91
  updating, 92
  using, 90–92
Tushman, M. L., 46

unions, support for skills training, 257
United Auto Workers, 257
U.S. government
  assistance for small manufacturers, 187–88, 256
  Department of Defense (DoD), 188
  Department of Education, 255
  Department of Energy (DOE), 187
user-centered diagrams, 111, 115, 116
utilities, quick-disconnect, 181

value chains, 117–18
value, of agile manufacturing, 266–72
Van De Ven, A. H., 46
vehicles, material handling, 148, 181
velvet glove mandate, 52–53, 55, 277–78
  at Xerox, 64
vendors. *See* suppliers
virtual manufacturing tools, 175
virtual organizations, xviii
visions, 192, 225, 226
  communicating, 52–53, 265
  developing, 51–52, 247–49
  implementing with strategies, 192
  for work groups, 265–66

visual production signals, 7, 10, 148–49, 172
  electronic, 182
Vocational Educational Act, 256

Wachovia Corporation, 246
Walt Disney Company, 247–48
Weisbord, Marvin, 97
"Why Change Programs Don't Produce Change" (Beer, Eisenstat, and Spector), 31–32
The Wisdom of Teams (Katzenbach and Smith), 98
workers. *See also* teams; training
  applying knowledge of, 244–45
  commitment of, 250
  cross-training, 158, 259
  direct customer contact, 237
  empowering, 159, 270–72, 277
  equipment monitoring by, 20
  expanded influence of, 107
  involvement in process redesign, 97–98, 250, 271
  literacy levels, 255–56
  preventive maintenance by, 7, 9, 16–17, 172
  quality control by, 7, 9, 11–12
  resistance to change, 278
  respect for, 254
work-in-process (WIP)
  in conventional manufacturing, 9
  costs of, 149–50
work processes. *See* processes
workstations. *See* shop floor
world-class manufacturing, 145

Xerox, 58–66

Zengler, J., 269